Japanese Marxist

Harvard East Asian Series 86
The East Asian Research Center at Harvard University administers research projects designed to further scholarly understanding of China, Japan, Korea, Vietnam, Inner Asia, and adjacent areas.

Japanese Marxist
A Portrait of Kawakami Hajime
1879-1946

Gail Lee Bernstein

Harvard University Press
Cambridge, Massachusetts
and London, England
1976

HX
412
K345
B47

Copyright © 1976 by the President and Fellows of Harvard College
All rights reserved
Preparation of this volume has been aided by a grant from the Ford Foundation.
Printed in the United States of America

Library of Congress Cataloging in Publication Data
Bernstein, Gail.
  Japanese Marxist.

  (Harvard East Asian series; 86)
  Bibliography: p.
  Includes index.
  1. Kawakami, Hajime, 1879-1946. I. Title. II. Series.
HX412.K345B47   335.43'092'4 [B]   76-20516
ISBN 0-674-47193-8

3 3001 00669 8285

*Dedicated to my mother, Edna L. Bernstein,
and to the memory of my father,
Bernard Bernstein (1907-1964)*

acknowledgments

So many American and Japanese mentors and friends helped me write this book that it seems impossible to thank them all adequately for their time, patience, and encouragement. In the limited space afforded here, I can only briefly express my gratitude to some of those persons and institutions on both sides of the Pacific Ocean who contributed in numerous important ways to the publication of the present work.

Albert M. Craig of Harvard University guided my research from its beginnings to its present form. His fastidious reading of the manuscript in its various stages, his relentless prodding, and his unflagging support accounts for the shaping of this study into a publishable monograph. To him I owe whatever I have learned about how to write a book.

It is hard to imagine how I could have accomplished any research in Japan without the assistance of Ishida Takeshi of the Institute of Social Science, University of Tokyo. By arranging for me to use his office and the facilities of the institute, and by introducing me to other members of the Japanese academic world, supplying me with necessary books and articles, and anticipating my research needs even before I did, he smoothed my way countless numbers of times. My debt to him is very great.

Among the other members of the Japanese academic community whom I should especially like to thank are Maruyama Masao, professor emeritus of Tokyo University, for a long and fascinating interview; and Ōuchi Hyōe, also professor emeritus of Tokyo University, and Furuta Hikaru, of Kamakura National University, both biographers of Kawakami, for kindly consenting to discuss my work with me. At Kyoto University, Hattori Masaaki graciously allowed me to share his office. Motoyama Yukihiko provided not only intellectual stimulation and practical assistance but he and his wife extended a warm hospitality that I deeply appreciated. Matsuo Takayoshi had a way of turning up valuable documents whose existence nobody else knew about: I cherished both his example of high academic standards and his gentle kindness. Sumiya Etsuji of Dōshisha University, another biographer of Kawakami, lent me his books and granted me an interview.

Bracing discussions with Peter Ch'en helped me formulate some of the main themes of this book. I also profited from the suggestions made by Peter Duus, Richard Minear, Benjamin I. Schwartz, Henry Dewitt Smith II, and Patricia G. Steinhoff. Suresht Renjen Bald, Charles H. Hedtke, Noriko Lippit, and J. M. Mahar read portions of the manuscript and offered valuable criticism. Tomone Matsumoto proved to be an indefatigable research assistant in the last stage of revision.

I am grateful to Ronald C. Miao for providing the calligraphy for the part pages. Yasue Aoki Kidd supplied the characters for the bibliography, and Jeffrey P. Wagner prepared the index.

Generous support from the Foreign Area Training Program between 1963 and 1965 enabled me to conduct research in Japan, and a position as research associate at the East Asian Research Center of Harvard University in the summer of 1973 greatly facilitated my progress in revising the original manuscript.

While living in Tokyo, I was treated to all the physical and emotional comforts of home by Mrs. Ishikawa Toyoko who, together with her family, attended to my needs with loving devotion. In Kyoto, Mrs. Muto Mieko welcomed me into her Japanese-style home, providing an ideal environment for my studies.

Michael Patrick Sullivan carefully read the entire manuscript. I relied greatly on his practical advice and moral support, and I presumed heavily on his friendship. He deserves a special word of thanks.

contents

PART ONE: MEIJI NATIONALIST
  1 A Young Man of Meiji Japan   3
  2 Crisis   19
  3 Meiji Dropout   34
  4 The Way in the Modern World   51
  5 Japan and the West   71
PART TWO: ACADEMIC MARXIST
  6 The Road to Marxism   87
  7 The Meaning of Marxism   103
  8 Historical Materialism and Revolutionary Will   117
PART THREE: COMMUNIST REVOLUTIONARY
  9 The Professor as Political Activist   131
  10 Working for the Communist Party   147
EPILOGUE: PRISON YEARS AND BEYOND   163
NOTES   177
BIBLIOGRAPHY   199
GLOSSARY   213
INDEX   217

Illustrations (page 84)

Courtesy of Chikuma shobō, from *Kawakami Hajime chosakushū*
Kawakami Hajime in 1906
   in 1924 with his parents, wife, and children
   in 1928 at a Labor-Farmer party convention
   in 1944

preface

One cannot fully comprehend twentieth century Japan without understanding the impact of Marxism on the intellectual framework and political culture of Japanese intellectuals: the chief opposition party in postwar Japan has been a Marxist socialist party; student radicals, militantly Marxist, have sought to influence Japanese politics through street demonstrations, university strikes, and propaganda posters; and Marxist ideals and economic analyses have often shaped the Japanese conceptualization of society.

Although many Japanese contributed to the dissemination of Marxism in Japan in the post-World War I period, I have chosen to write this intellectual biography of Kawakami Hajime (1879-1946), one of the founders of Japanese Marxism, because as the son of a samurai family, born only a decade after the feudal order was dismantled, Kawakami embodied many of the tensions between Japan and the West that have haunted the Japanese nation. I was drawn by Kawakami's heroic and ultimately tragic struggle to reconcile his personal, moral, and religious needs—the Japanese side of his being—with what he felt to be the scientific truths of Marxism. This struggle was, I feel, characteristic of Japanese intellectuals' efforts throughout the last one hundred years of Western cultural influence and breathtaking change to achieve the standards of modernity while preserving their own cultural identity.

Kawakami began to study Marxism in 1919, at the age of forty. A professor of economics at prestigious Kyoto Imperial University, he influenced an entire generation of post-World War I students, who were inspired by his lectures to study Marxism and to become involved in political causes. Kawakami's books and his writings in the popular press helped disseminate Marxism to a wide audience of intellectuals. To know the circumstances of his conversion to Marxism is to know some of the dominant intellectual and political problems of his day. Kawakami's interpretations of Marxism and his debates with other Marxists reveal the terms in which reformers' dialogues were cast. Why was he attracted to Marxism? What was his understanding of Marxism? What solution did he see in Marxism to the social problems confronting Japanese society? And, above all, why did he hesitate to commit himself fully to Marxist teachings?

This book seeks to answer these questions by describing the gradual evolution of Kawakami's thought from patriotic nationalism to humanistic socialism and, finally, to revolutionary communism. His evolution as a Marxist closely paralleled the history of economic thought in Japan. A pioneer in the little-known field of economics in the late Meiji period, Kawakami Hajime's career as an authority on German nationalist economics, British classical economics, German Marxism, and Russian communism formed part of the larger history of Japanese efforts to comprehend the economic principles governing industrialization.

Kawakami was primarily attracted to Marxism as a solution to the problem of poverty. He viewed Marxism as a science of the distribution of wealth and saw in it a method of continuing the economic development begun under the capitalist system while eliminating the two related evils of poverty and selfishness. Early in his career as a student of economics, Kawakami had tried to modify the individualist, selfish ethos of classical economics in order to make it ethically acceptable. Drawing on Confucian, Buddhist, and Christian teachings, he formulated his own syncretic ethics of absolute selflessness to guide the nation's morals and to cure its social ills. Failing in this attempt to apply moral remedies to social problems, Kawakami turned to Marxist solutions.

Ironically, Marxism forced Kawakami to abandon his commitment to absolute standards of morality and to think in entirely new categories of thought: class war, class ideology, the dialectic. Marxism, in its Leninist form, moreover, also forced him to abandon his scholarly detachment in favor of revolutionary practice: in 1932 he joined the Japan Communist Party. But he learned these lessons in modern politics and ideology only with great difficulty.

Kawakami's difficulties with Marxism stemmed from his traditionalist approach to ethics and politics. Japanese scholars generally agree that Kawakami's unique political style as a Marxist derived from his samurai background and his early Meiji education in statecraft. Kawakami's devotion to nation and society, his view of politics as a moral vocation, and his urge to self-sacrifice on behalf of a cause higher than himself were ideals motivating not only the patriots of the Restoration period but, as Kawakami's later life and thought demonstrated, such ideals continued to drive revolutionary actors in later decades of Japan's modern history as well.

It was precisely the intrusion of these Japanese values into his Marxist studies, however, that earned him the rebuke of more objective Japanese Marxists. Kawakami's development as a Marxist became so entangled with his ethical and religious concerns that the two can be distinguished only by doing serious damage to our comprehension of his thought. Both Kawakami and his critics understood this point quite well. The conflicts between ethics and politics, religion and science, and, ultimately, Japan and the West, became the leading themes of Kawakami's life. Unable to resolve these conflicts, Kawakami concluded toward the end of his life that he was a special Marxist. What interests me about Kawakami is his ambivalent attitude toward Marxism and toward the Western culture that spawned it, an ambivalence that haunted him throughout his adult life.

Depending upon their own ideological position and their distance from Japanese tradition, the many critics, friends, relatives, and former students of Kawakami Hajime who have attempted to explain his thought either speak warmly of him as an anomaly, a Japanese Marxist in search of the Way, or scorn him for being an imperfect Marxist. Kawakami's autobiography and memoirs went far in establishing his own interpretation of his life as a spiritual quest. If there is a legend of Kawakami Hajime in Japan, it is in no small measure because he successfully dramatized the story of his own life, casting it in terms of the pursuit of selflessness. Long after he lost his position as the preeminent interpreter of Marxism, his fame as a Marxist was kept alive by the extraordinary human interest of his spiritual life history.

When writing about the life and thought of someone who is no longer alive and whom one has never met, it is impossible to escape a great degree of dependence upon the subject's own evaluation of himself. Most of our information on Kawakami Hajime's life comes from his five-volume autobiography, written after his prison sentence in the nineteen thirties and published following the end of World War II, when it became a bestseller. Almost everyone who has written or spoken about Kawakami in the post-World War II period has relied uncritically on the autobiography. Although it is an invaluable guide to Kawakami's personality, it must be treated not as an exact image of the man, but as a portrait of the man he wanted his many readers, at the end of his life, to remember. I have wondered in places whether Kawakami "really felt that way," but as far as the historian is concerned, what Kawakami said he thought or felt is as much a fact as what, subconsciously or otherwise, he really did feel. However, when at the age of sixty, he describes his mental state at the age of twenty, there is need for a full measure of caution. Here I have tried to corroborate accounts in Kawakami's autobiography with articles and letters written during the specific period in question. A literary extrovert, Kawakami shared his opinions and feelings fully enough with the public at large to leave behind a voluminous record of his inner life history. This record reveals a talented, dedicated, emotional, fallible, self-righteous, and thoroughly human personality.

One rarely finds studies of Japanese in English which disclose something of their personalities. Numerous political biographies have been published but few books convey a sense of the interior life of their subjects. The heir of a samurai family, an authority on economics, a professor at one of Japan's foremost Imperial Universities, an early popularizer of Marxism in Japan, and a Japanese Communist, Kawakami Hajime is important enough in the modern history of Japan to merit a study of his ideas. My own motivation for the present study, however, is to provide a glimpse into the inner life of a Japanese who lived through the upheaval of Japan's modern century and wrote about it with candor and color.

part 1. meiji nationalist

1. a young man of meiji japan

Sprawling across the southwestern portion of the main island of Japan, Chōshū in the middle of the nineteenth century was one of the most powerful of Japan's three hundred feudal domains. It was also one of the first to feel the shock of Western penetration. Commodore Perry's arrival in Japan in 1853 turned Chōshū into a center of political intrigue, where rebel groups hatched daring plots—to stow aboard foreign ships, to sneak off to Shanghai and London, to murder conservative Tokugawa officials who appeared willing to compromise with the enemy's treaty demands.

Some of these defiant shishi, as the bold samurai patriots were called, were arrested and executed or ordered to commit suicide for their acts of disobedience, but a few succeeded in rising to leadership positions within the domain. Recognizing the futility of driving out the foreigners unless the country was united, they planned instead to overthrow the Tokugawa House, whose divisive policies had left the country vulnerable to foreign encroachment. The young patriots organized a peasant militia and, in alliance with rebel domains, they won the support of the fifteen-year old emperor for a march on the Tokugawa stronghold in northeast Japan. Calling for a restoration of imperial rule, the Chōshū loyalist samurai forced the Shogun's surrender and in 1868 they seized control of Japan. The Meiji Restoration of 1868 marks the beginning of the modern Japanese nation.[1]

One of the soldiers who fought against the Tokugawa armies was Kawakami Sunao (1848–1928), a retainer for the Kikkawa family, lords of Iwakuni, a branch of Chōshū. The Kawakami family lived in the small village of Nishikimi, across the river from the castle town of Iwakuni and directly south of Hagi, where the main house of Chōshū, the Mōri, resided. The Kikkawa and the Mōri were related, and although ties between the two daimyo families had been strained since the beginning of the Tokugawa period, they had settled the grievances between them ten years after Perry's arrival and joined forces in 1868 to help defeat Shogunal rule.[2] Kawakami Sunao had marched under the Kikkawa banner in the 1868 Restoration,[3] embracing the Chōshū cause, and after the establishment of the new imperial government, he had become village chief of Nishikimi, where Kawakami Hajime, his first child, was born on October 20, 1879.[4]

Samurai Son

Growing up as the son of a Chōshū samurai family, Kawakami Hajime's future was inevitably tied to the major reforms that were transforming Japan from a feudal society into a modern nation. Such illustrious Meiji statesmen as Itō Hirobumi, architect of the modern Japanese constitution, and Yamagata Aritomo, founder of the modern Japanese Army, were born in Chōshū and grew to manhood in the turbulent years after the West forced open Japan's ports to trade and diplomatic relations. Together with their Chōshū protégés and coconspirators from Satsuma, they created during the 1880s an oligarchic form of rule which would dominate politics in Japan into the early twentieth century. The oligarchs were committed to the task of building a strong, unified Japan, powerful enough to resist further foreign encroachments and modern enough to assume a prominent position on the international stage. The very name Chōshū was freighted with the heady connotation of national leadership and political power. Reared in the birthplace of these shishi activists and nurtured on the romance of Restoration politics, Kawakami Hajime could not fail to be influenced in his own lifetime by the ideals of statecraft which his clansmen personified.

The Kawakami family's lower samurai status and modest income placed them, at the end of the Tokugawa period, in much the same social rank and desperate economic circumstances as the future leaders of Meiji Japan. Hajime's grandparents had barely survived on their meagre stipend of nineteen koku,[5] an amount considerably less than the one hundred koku deemed adequate to support a family comfortably in the late Tokugawa period.[6] To make matters worse, when Kawakami Hajime's father, Sunao, was five years old, his own father, Sai'ichirō, died, leaving the family almost destitute. Since Sunao could not qualify for his father's stipend until he himself reached the age of twenty, his mother, who was only twenty-five years old when she was widowed and who never remarried, was forced to work at menial tasks in order to pay her husband's debts and raise Sunao and his three-year old sister.[7]

By the time Hajime was born, however, the family's circumstances had improved. Although Sunao was deprived of his samurai stipend in 1869, when feudal classes were abolished, the income he received from his office as village chief shielded him from the economic distress afflicting other members of the old elite class. Much as Kawakami Hajime insisted on his humble origins, the fact remains that his father's salary was more than adequate for the family's needs. The family was not rich, but they were locally prominent, for Sunao's position in the bureaucracy, even at his low administrative level, provided him and his family with the respect traditionally reserved for bureaucratic officials.

Sunao served as village chief until 1889 and then, when Nishikimi merged with Iwakuni, he was elected the first head of the new township. Well known and admired by the ten thousand residents of his administrative area, Sunao embodied the virtues of loyalty and moral rectitude long considered synonymous with samurai class station, and he became his son's earliest example of devotion to public office. In all his years in office, he was never involved in any scandals nor suspected of profiting personally from his political influence. "In today's world," Hajime's father-in-law once said, "there is nobody as honest as Kawakami Sunao."[8]

Sunao was thirty years old when he married Kawakami Tazu, the sixteen-year old youngest child of a samurai family distantly related to his own ancestral line.[9] He was, in a sense, marrying up. Tazu's father, Matasaburō, a private secretary in the employ of Kikkawa, had enjoyed a slightly higher status than Kawakami Sunao and had received a stipend of eight koku more.[10] Nevertheless, for reasons that are not known, Sunao divorced Tazu nine months after the marriage. Conflict between her and her mother-in-law may have been responsible for the divorce, because we know that the relationship between a man's wife and his mother was often critical in determining whether the marriage would succeed.

Four months after the couple was divorced, Sunao remarried, but learning shortly afterward that Tazu had given birth to a son, named Hajime, he arrived at her house, dressed in formal clothes and accompanied by a wet nurse, to take the infant home with him.[11] Tazu apparently yielded without protest to her former husband's wishes: in traditional society it had been customary for divorced women to return without their children to their parents' home. The task of raising Hajime thus fell to the wet nurse and the stepmother, under the general supervision of Sunao's widowed mother.

Almost immediately after the infant arrived at his father's home, however, problems arose over his care. The wet nurse was soon dismissed after Hajime's grandmother discovered that she was having difficulties breastfeeding and suffered from asthma. The stepmother was sent away about one year later when, shortly after the birth of her own son, she punished Hajime by hanging him by one arm down a well.[12] When Hajime was two years old, his real mother was brought back into the household but, busy with domestic chores and too timid to assert herself, Tazu left the care of her son to the willing hands of her mother-in-law, while she herself assumed responsibility for raising Nobusuke, her husband's one-year old son by his second marriage. A third son, Sakyō, was born several years later. In the few passages in his memoirs where he mentioned his mother, Kawakami usually described her performing household tasks, such as bringing trays of sake to his grandmother's quarters.

The major influences in the boy's early upbringing were his father and especially his grandmother, Iwa, from whom Hajime inherited many aspects of his physical make-up. His prominent cheekbones, narrow face, and somewhat full lips were more like those of his grandmother than of his mother, who was rather plump, with a round face and softer features. A picture of the three of them taken when Hajime was about three shows him resting against his grandmother's knee, as she sits holding one of his hands in her lap, while his mother, only about twenty at the time, stands demurely to one side.[13] The picture symbolizes the relative roles played by these two women in Hajime's early life.

Grandmother Iwa was fifty-two years old when Hajime was born, and with her daughter-in-law to serve her, she now had leisure time and some savings to lavish on her eldest grandson. By dint of hard work, she had managed to acquire enough money to build separate quarters for herself attached to the original family house. Sunao, his wife, and his second son lived in the old house, while Hajime slept in Iwa's room. This arrangement was common practice among Japanese families; especially after the birth of a second child, it was not unusual for the first-born to become "grandmother's child." It was also common practice for the paternal grandmother to dominate the household and here, too, the Kawakami family was no exception.

Iwa doted on her grandchild. She picked him up as soon as he started to cry and walked back and forth in the garden to comfort him, with Tazu holding the lantern to light the way. When he was still learning to crawl, she scattered sweet cakes on the floor to encourage him, and after Hajime began going to school, his father arranged for the school janitor to carry the little boy on his back, while Hajime's younger half brother trotted alongside. Wherever Iwa went—to the theater, on picnics, sightseeing at local shrines or Buddhist temples—she took her grandson along.[14]

Hajime's grandmother was strongwilled, independent, and eccentric, qualities which Hajime absorbed. She was determined to enjoy the remaining years of her life and did so in ways that deviated from the norms of samurai behavior. Ignoring the gossip of villagers, she took a young lover to live with her and her grandson in the attached house. She drank sake daily, often becoming drunk and loud, and she openly attended risqué theatrical performances, accompanied by Hajime and her lover. Eventually she arranged a marriage for her lover and, as if to defy community standards even further, she appeared at the wedding celebration.[15] Although Iwa's behavior was unusual, to say the least,

Sunao never once rebuked her. Perhaps out of sympathy with her many years of work and her emotional deprivation, he not only overlooked her activities, but he followed her wishes in regard to his own family life.

It is difficult to evaluate the effect, if any, of Iwa's morals on Hajime's psyche. He witnessed her heavy drinking, and he heard her flirtatious banter. Since he slept in her room as a child, he may have also been exposed to the more intimate aspects of her relationship with her lover. If these experiences left psychological scars, Kawakami never even hinted as much in his autobiography. Rather, his memories of childhood evoke mellow times, when he was enveloped in the warmth of familial love, treated with special respect as the eldest son of the village chief, and endlessly entertained by Iwa, who proudly escorted him across the picturesque Bridge of the Brocade Sash over the Iwakuni River to visit friends and relatives in nearby towns or to see the colorful festivities and daimyo pageants that enlivened the slow rhythm of village life and brought all the households together in a sense of shared community. Kawakami's adult vision, unlike that of many other reformers, was surely not "intimately connected with the wrongs of childhood."[16]

Yet, one cannot help but note in the autobiography symptoms of anxiety and anger in the young Hajime. Around the age of three he was given to temper tantrums and screaming fits of such unmanageable proportions that his grandmother, afraid to leave him alone and unable to pacify him, tied him to a pillar in her room when she went to take her bath. Upon occasions his violent rages forced his mother and grandmother to hide in the family storehouse, hoping to escape from his wrath, but Hajime would stand outside, pounding on the door.[17]

Although we lack systematic studies of childrearing patterns among samurai families in the early Meiji period, we do know that both Mori Arinori and Yamagata Aritomo, samurai of a generation immediately preceding Hajime's, were subjected from an early age to strong doses of discipline. The training was administered by female members of the family—mothers and grandmothers—who used the male household heads as models for the deportment expected of the younger males in the family. Their method of inculcating manners and the steely virtues of courage and stoicism were often more persuasive than coercive: parables replaced the rod.[18] The female members of Kawakami's family, however, appear to have relaxed the standards in Hajime's case. "From the time I was a child, I had a terrible temper. I would get angry at something, burst out crying, and become completely uncontrollable."[19] It is true that eldest sons in samurai families were traditionally favored above other children in the family and were often treated with special regard; nevertheless, Hajime's early years, according to his own recollections, were noteworthy for the exceptionally free rein he was given, often at the expense of other household members. "After I grew up, I often heard stories about how my grandmother, in particular, had pampered me, and about how I, in particular, had given her a lot of trouble."[20]

Hajime's frail physical constitution may explain the reluctance of family members to upset him by imposing disciplinary measures. Throughout his life, he suffered from a sensitive digestive system and a tendency toward psychosomatic symptoms, such as stomach pains and fevers, in times of distress. From infancy on he was a delicate child and, despite strenuous efforts of family members to tempt him with special foods, he never ate enough to fill out his scrawny body. He suffered from a number of ailments so serious that, when he was three years old, doctors predicted he would not live beyond childhood, and when he was around twenty years old, he was still too sickly to qualify for a life insurance policy.[21] This burden of chronic illness must have contributed to Hajime's temperamental nature; and by their worried concern, Hajime's maternal and parternal grandparents further encouraged his emotionalism.

There are several other possible reasons for the rage and nervous instability evidenced in Hajime's behavior during his early years. These traits of character are worth discussing, as they had some bearing on Hajime's adult personality. One reason for what he called his "horrendous early childhood willfulness" may be related to the trauma of losing his real mother, mistreatment by his stepmother, and the general confusion of surrogate mothers that characterized the first three years of his life. Not knowing what was considered normal in this period, it is hard to assess the psychological impact on Japanese children of being reared by women other than their own mothers. Divorce, adoption, and the general shuffling of wives and children back and forth from one family to another may well have been fairly common events in a society that viewed the continuity of the family line as more important than the needs or wishes of individual members. Sunao's sister was divorced from her first husband on the grounds of infertility; she remarried and was divorced a second time, only to remarry her first husband again. Several well-known figures of the Meiji period, such as Itō Hirobumi and the novelist Natsume Sōseki, were adopted into other families. Sōseki's severe mental disturbances and Kawakami Hajime's harrowing experiences with his stepmother, however, illustrate some of the possible terrors awaiting an unwanted child. In his autobiography, Kawakami dwelled at some length and with a tone of mild resentment on the fact that he was not breast-fed, blaming his physical frailty, his cavities, and his weak stomach on this nutritional deficiency and pointing out that his brother, Sakyō, who had been breast-fed, was stronger. Even as an infant, Hajime evidently suffered from the loss of maternal comfort, for he tried unsuccessfully to nurse at his grandmother's breast.[22]

A second possible factor influencing Kawakami's behavior was the streak of epilepsy that ran in both sides of the family. In her youth, Iwa was notorious for sudden outbursts of rage which grew more infrequent as she grew older. Later in his own life, Kawakami realized that the fits he remembered seeing as a child were neither temper tantrums nor hysteria, as they were then diagnosed, but epileptic attacks, and he suspected that he may have inherited the illness.[23]

In any case, Iwa's own behavior was hardly calculated to inspire her grandson's timid acquiescence in conventional standards. She and Sunao seemed eager to tolerate and even encourage the fragile child's assertiveness. Independence and self-reliance were becoming prized qualities in Meiji society, and perhaps Kawakami's family believed that a strong will would enable Hajime to succeed in the highly competitive environment of the new social order. Such a belief was once voiced by a tutor hired to help Hajime, then around nine years old, with his studies. Exasperated by the boy's restlessness and his tears of frustration when made to sit still, the tutor told Hajime's parents, "He will surely be successful when he grows up, since he hates to be defeated."[24] Incidents such as these, narrated by family members and remembered by Kawakami many years later, must surely have helped shape and reinforce his own self-conception.

If the Kawakami family did indeed wish to strengthen young Hajime's will, they succeeded. Kawakami's brother-in-law described him as a man of strong will and iron determination who, once his mind was set, could not be deflected from his chosen course of action.[25] His elder daughter painted a less flattering and strikingly unfilial portrait of her father as a spoiled brat (*dadakko*) with a violent temper who never outgrew his demands that family members cater to his needs.[26] Perhaps the major difference between the willful child and the equally self-centered man of his daughter's memory was that the adult Kawakami made demands in the name of his life work, and the anger he had once directed against parents and grandparents was turned against social injustice.

Sunao provided little direct help in taming his son's stormy personality; his influence, though benign, was remote. Hajime recalled enjoying a warm relationship with his father, even though the two could hardly be more different in temperament. Sunao was a model of self-control and dignity, a man whom his nephew could not imagine being "in any posture other than bolt upright."[27] He was not an authoritarian figure;

rather, as the ideograph of his name suggests, he was mild mannered and calm, with a gentle, even passive demeanor that his son respected but was never quite able to emulate. "If you want to express my father's character and his whole life in one word," Kawakami wrote, "no other word is more appropriate than 'Sunao.' "[28] After his father's death, he chose a posthumous Buddhist name for him that connoted one who had reached the other shore of religious enlightenment, and he wrote with evident fondness, "Fortunately, I do not have one bad memory of my father."[29]

Kawakami Sunao's example of prudent behavior and constancy were appealing but elusive qualities for the headstrong and mercurial Hajime who, unlike his father, was at various times in his life given to impulsive and even rash acts. Hajime frequently agonized over decisions, changed his career plans half a dozen times before settling down and never lost the restless urge that caused him to leap, as he put it, into noble causes. There was little of Sunao's docility in his first son.

We can attribute some of these marked contrasts between father and son simply to the vagaries of genetic endowment and also to the considerable differences in their early home life. But beyond these two factors, we can perhaps see in such extreme personality differences the imprint of the changing historical times. Sunao was essentially a conservative, whose ethical ideals exemplified the traditional samurai code of loyalty to political superiors. Retiring from village office in 1905, a few years short of his sixtieth birthday, he worked for the next twenty years as a steward for the Kikkawa family, walking every day across the Iwakuni River to the old castle until shortly before his death at the age of eighty. A healthy man, with good coloring and a radiant smile, Sunao seemed content to live out his days in his native village, dutifully serving the Iwakuni lords, his life little affected by the drastic reforms that were sweeping across the nation. "Unlike me," his son wrote, "he hated change."[30]

The society into which Hajime was born, however, was in upheaval, and it was difficult to prepare any boy, much less the high-strung Kawakami heir, for an uncertain future. Much as Hajime admired his father, his own explosive nature and the turbulent times in which he gew up eventually created in him the need for new models to guide his behavior. It may be significant in this regard that Sunao has been described as pliant and yielding almost to a fault: he got along well with others, according to Hajime's brother-in-law, by overlooking their misdeeds. Otsuka Yūshō recalled that, although Sunao was scrupulously honest in his own behavior, he was also naive; and he failed to see dishonesty in others, whereas both Hajime and his brother, Sakyō, an artist, were more demanding of people and less tolerant of injustice.[31] The traits of character that typified Kawakami Hajime's adult personality, as described by those who knew him well—his inflexibility, his intensity, his hot temper—represented the very opposite of the man his father was. Kawakami's lifelong preoccupation with an ethics of complete selflessness is interesting in the light of his upbringing, for if it is true that early family relationships have some bearing on later intellectual convictions, then Kawakami Hajime may have been reacting not against parental authority, but against the lack of it.

In comparing father and son, it may also be significant that Hajime himself was a stern, authoritarian father, lacking in the affectionate warmth his own family had showered on him as a child. "From early times he had a strong temper," his daughter wrote, "and when he said 'This is the way it's going to be,' no matter how patently unreasonable it was, he would insist on being obeyed." Kawakami remained aloof from family matter, spent little time with his children, and rarely indulged in displays of affection. "My father seldom held his children," Kawakami Shizuko recalled.[32]

This is not to say that Kawakami Hajime was emotionally callous. The major differences in the behavior of Sunao and Hajime as fathers quite possibly stemmed from the greater pressures placed on Hajime to excel in his professional life. Sunao's phlegmatism reflected the security of one who had essentially inherited his position in society under the patronage of more influential political leaders; Hajime's intensity betrayed the tensions of competing for his own place in a new society.

But Hajime learned about this society only after his formal schooling began. Over the years he came to understand that beyond Nishikimi lay a whole new world where challenge was the order of the day. Neither Sunao, who hated change, nor Iwa, who loved pleasure, could protect him from the turmoil of growing up in early Meiji Japan.

The Making of a Meiji Statesman

After the abolition of feudal privileges in 1869, education rather than inherited rank became the single most important determinant of social status in Meiji Japan. The new government desperately needed able men to manage their crash program of political and economic reform, a program designed to make Japan an equal of the Western powers. Opportunities for rapid advancement within the national bureaucracy beckoned talented and ambitious young men.

Recognizing the value of schooling in Western subjects, and determined to preserve their domain's commanding position in national politics, the former lords of Chōshū took the initiative in establishing modern educational facilities in the early years of the Meiji period. The school system they created in Yamaguchi Prefecture (the new administrative unit embracing the former Chōshū territory) was perhaps the most advanced in all of Japan. Under the urging of Inoue Kaoru, Minister of Foreign Affairs and a native of Chōshū, the Mōri family and the lords of their former branch han funded a prefectural school system, known as the Bōchō Educational Association (*Bōchō kyōiku kai*), to teach the Western learning so necessary for their young students' successful competition for government office. Children of low-ranking samurai, despite their families' straitened financial circumstances, in this way gained access to the special training in practical learning required of them as the nation's future leaders.[33]

The Kikkawa family became pioneers in the development of modern schools in the Iwakuni area. Early in the Meiji period, they converted the old fief's military school into a public middle school (*chūgakkō*), and the school of literary arts (*bungakkō*) into a public primary school (*shōgakkō*) that became a model for the entire prefecture. Appreciating the importance of foreign language training, they also established a language school in 1871, and between 1872 and 1874 they hired an Englishman named Herbert Stevens to teach English as well as science and arithmetic courses. In keeping with the Meiji government's enlightened policies, the school even offered special classes for women.[34]

Hajime began his formal education in Nishikimi at the age of four and a half, about one and a half years younger than the average first grader. His father's duties extended to headship of the primary school, a fact that may explain why Hajime was admitted ahead of his age group and also why his teachers consistently gave him high grades on his report cards, even though he hated school and remembered being no more than an average pupil. As a result of his father's "foolish haste," Hajime later failed to be promoted at the end of his first year in his next school, in Iwakuni, and required private tutoring to prepare him for advancement to the next grade.[35] Through this unpleasant experience, he learned for the first time that family influence alone was not enough to succeed.

The change that came over Hajime in the years following his academic failure was sudden and striking. He grew from a little boy who cried because his tutor would not let him play in the garden into a young man with a love of reading and writing and impressive leadership qualities. His education in Iwakuni must have played no small part in effecting this transformation, for the school curriculum in the former castle town was designed to socialize the sons of Chōshū samurai into the responsibilities of statecraft.

The Iwakuni School, one of five schools established by the Bōchō Educational Association, enjoyed the reputation of being the starting line for the race to the top of the bureaucratic and academic worlds.[36] Young Kawakami's five years of schooling in Iwakuni, from the age of eight and a half to thirteen and a half, were instrumental in developing his patriotic consciousness and political ambition, imbuing him with an elitist self-image and a sense of mission that directed his willfulness into socially constructive channels. The school curriculum fostered these sentiments by supplementing courses in Western learning—German and English languages and science—with the old han's traditional education in Confucian statecraft. Students were raised on stories about the heroism of the sons of Chōshū, who had sacrificed themselves for the nation, and they memorized the words of the famous Chōshū loyalist, Yoshida Shōin, who had defined the ideal man as the shishi. The patriotic "man of will," Shōin wrote, "took the nation's weal as his own."[37]

Political ambition and patriotic idealism were thus inextricably combined in the education of young Chōshū students. They were taught that government office was a moral vocation—their duty as an elite, fitted as they were for national leadership by their ancestry, their education, and their patriotic fervor. Although young men were encouraged to assert themselves aggressively, imitating the spirit of individual self-reliance that presumably accounted for the strength of the "civilized" nations of the West, they were expected to devote their energies to the fulfillment of Japan's needs. The curriculum encouraged students to rise in the world (*risshin shusse*) not for their own reward, but for the greater glory of Japan. Hajime's determination to make his will prevail was a trait which, carefully guided, served the interests of Japan's polity.

It is remarkable how well the young men of Meiji Japan learned their lessons. Their early political awareness and elite consciousness created an entire class of teenage intelligentsia, linked to Tokyo, the center of political life many hundreds of miles away, by only newspaper and journal subscriptions. If they were at all like Kawakami Hajime, they were avid readers. Even in early adolescence, young Kawakami was already in touch with political and literary trends emanating from the nation's capital. Every night before going to bed, he read copies of Fukuzawa Yukichi's *Current Affairs (Jiji shimpō)*, sent to him from Tokyo by his uncle. He subscribed to another magazine called *Child's Garden (Shōnen-en)* and, together with some friends, put out a weekly magazine of his own, entitled *Report (Kaihō)*.[38]

Stimulated by his new environment, Hajime quickly made up for his earlier academic failure by demonstrating a precocious flair for history and politics. Together with a friend, he voluntarily requested instruction from a local teacher, who helped the boys read about the heroes of Japanese history in Rai San'yō's *Unofficial History of Japan (Nihon gaishi)*. Hajime also introduced the custom of using the school's rice-polishing room to practice public speaking with fellow classmates in imitation of the fashionable political debates and speeches sponsored by the Popular Rights movement. These *enzetsukai*, which Iwa had taken Hajime to hear, were introduced by the famous educator Fukuzawa Yukichi as a way to develop individual initiative, and their popularity was an example of the Meiji craze for civilization and enlightenment. A few months before his eleventh birthday in 1890, Hajime once again demonstrated his political awareness by writing an essay expounding the familiar *fukoku kyōhei* slogan about the nation's need for industries and a modern army.[39] By the time he wrote this essay, he recalled, "the seeds of a practical, political nature planted in me had already begun to germinate."[40]

In 1893, when he was fourteen years old, Hajime graduated from the Iwakuni school and left his home for the first time to continue his education in Yamaguchi City. Yamaguchi Middle and High School (*kōtō-chūgakkō*), also funded by the Bōchō Education Association, represented one further step in the ladder leading to the higher reaches of the national bureaucracy. It offered a three year preparatory curriculum of the liberal arts sort followed by two years of professional training to prepare graduates for admission to the Imperial University in Tokyo.

Hajime approached this new educational challenge with some trepidation. Yamaguchi City, the prefectural capital, is today only about fifty miles due west of Iwakuni, but Hajime's trip was considerably longer, because the two cities were not yet linked by rail service. He was required to travel by boat, boarding at a newly constructed port facility about one mile from home and changing to another boat part way through the journey. The last part of his trip, between the Inland Sea and Yamaguchi, five miles to the north, was once again overland. The length and difficulty of the trip meant that Hajime had to live in Yamaguchi for the school year. His entrance into preparatory school thus represented an important step on his road to adulthood, an event whose significance was felt by all members of the family. His uncle sent him a nickel-plated wristwatch as a graduation present, but Iwa and Hajime's four year old brother, Sakyō, less jubilant, cried on the day Hajime left.

By the time he reached his destination, after travelling all day in third class accommodations, Hajime was exhausted and miserable. He moved immediately into a boarding house approved by the school, but not even the presence of other students, who were, he noted, curiously cheerful, could cure his homesickness. In the absence of parental supervision, Hajime was expected to report regularly to one of his father's old friends, a school librarian named Kohara. Sunao sent money to Hajime through Kohara, whose wife never failed to greet Hajime politely with, "Your parents must be worried," words which inevitably brought tears to Hajime's eyes.[41]

Before long, however, Hajime adjusted to his new environment. Some of his classmates were friends of his from Iwakuni school days, and one of these young men agreed to share a room with him in Kohara's guest house. Although Hajime's family had always worried about his poor eating habits, his appetite improved after he moved to the Kohara residence. Meals were generous, especially when compared to contemporary boarding school standards. Hajime and his roommate ate three meals a day, including such special fare as sukiyaki, liberally sprinkled with sugar, and a special sweet cake sold in the city. Writing of his high school days from his jail cell some forty years later, he devoted several pages to memories of food in the Yamaguchi period.[42]

Yamaguchi school years were important in developing the latent artistic and literary talents that contributed much to Kawakami's future course as a literatus. Although he continued to follow national political events through a subscription to the *Yomiuri* newspaper, he pursued literary trends with an even greater fervor. It was in these years that he became acquainted with some of the classics of Japanese literature, such as *Tale of Genji* and the historical romances of the feudal period, but for the most part his tastes were modern. He liked the novels and poetry of such representative Romantic writers as Shimazaki Tōson and Yosano Tekkan. Their works frequently appeared in the monthly issues of *Literary World* (*Bungakkai*), a newly established magazine published by a group of young Christians under the leadership of Tōson and Kitamura Tōkoku. Among the other contemporary writers whom Hajime enjoyed reading were Kunikida Doppo, Tayama Katai, and Doi Bansui; and he became engrossed in Higuchi Ichiyō's "Growing Up" (*Takekurabe*), when it first appeared in serial form in 1896.[43]

Hajime's favorite writers shared an interest in exploring the interior life. Whether expressing passion, melancholy, or the sadness of growing up, they sought to portray the new, liberated Meiji individual's complex struggle to understand himself and his changing environment. These writers were also interested in experimenting with new literary styles, often of Western inspiration. At a time when writers of fiction were still open to the charge of being delicate and unmanly, they professed a belief in the artistic validity of subjective feelings, and they worked to establish the legitimacy of creative writing as a career.

The goals and accomplishments of contemporary poets and novelists, some of whom were only in their early twenties, planted in Hajime's mind the possibility of other career choices. He began to consider himself a budding writer and formed a *waka* poetry society and literary magazine among his school peers. To celebrate the literary side of his nature, Hajime asked his friends to call him by the nom de plume Fugetsu. He devised this name by combining the ideograph for maple tree—signifying October, the month in which he was born—with the ideograph for the last part of the name of a well-known writer of the day, Ōmachi keigetsu.[44]

Hajime's literary aspirations were encouraged by Yamaguchi school officials. Responding to the problem of growing unemployment among the many young men who had studied law as an entree into government service, they had begun to direct their students into other fields of study. Hajime's interest in literature was further stimulated by a book entitled *On Behalf of the Mother Tongue (Kokugo no tame ni)*, written by Inoue Kowashi, the Chōshū-born Minister of Education. In this book, Inoue, who was killed in the same year that Hajime entered high school, extolled the virtues of the native Japanese language. Inspired, Hajime nonetheless hesitated to commit himself to a career in literature.[45]

The question of whether to major in law or literature produced the first of many dilemmas in Hajime's life. Many bright young men in the middle of the Meiji period suffered from the same conflict between a literary career and the lure of government service. In an age overwhelmingly characterized by a political, practical, and utilitarian spirit, Meiji writers and artists found it difficult to pursue their craft without feeling out of step with the times. Exposed to the politically charged atmosphere of the Chōshū curriculum since he was eight years old, it was only natural that Hajime would now hesitate to devote himself to a literary career. He had been trained after all for service to the nation in the finest traditions of his birthplace. To ignore the call of duty and engage in frivolous artistic pursuits violated the canon of Confucian nationalist teachings.

Yet, Hajime was clearly endowed with literary talent and artistic sensibility. He enjoyed sketching landscape scenes and writing poems—creative activities that had been encouraged by his maternal grandfather, who was an artist, calligrapher, and traditional poet of some renown. Political ambition, realized in service to the nation, and literary creativity, expressed through poetry, were thus intertwined in Hajime's adolescent years, producing a tension that was never completely resolved. The romantic urge to self-expression, though submerged, reappeared later in his life, contributing to his unique approach to politics. This literary romanticism also helped contribute to his fame as a Marxist. Kawakami Hajime's autobiography, reminescences, and prison poetry, as well as the poems he wrote to celebrate each landmark on his road to Marxism, form an important, if not central, part of the corpus of his publications. With good reason, Hajime's high school teachers urged him to favor the artistic side of his personality. To those who knew him as a young man Hajime's physical frailty and emotional sensitivity seemed unsuited to the pressures of the political world.

Choosing sentiment over statecraft, Hajime for a time decided to specialize in literature. Observing a tradition of school boys in Japan that has survived to this day, he journeyed to nearby Miyajima shrine, purchased a small slip of paper on which prayers may be written, wrote down his wish—to enter the literature department—and entrusted it to the shrine. But he could not shake off the attraction of politics and almost immediately suffered second thoughts.[46]

The most important source of support for Kawakami's political ambitions in these late adolescent years was the Chōshū loyalist Yoshida Shōin, in honor of whom he had chosen Baiin as his second pen name. The name was written with the ideographs for plum tree—a reference to a tree in the Kawakami family garden—and the second part of Shōin's name. Although the sixteen year old Hajime, according to one of his roommates, read literary works almost exclusively,[47] he also regularly visited the shrine in Yamaguchi City dedicated to his childhood hero and pasted copies of Shōin's writings on the walls of his room.[48]

The aspect of Yoshida Shōin's life most familiar to the Western world is probably his unsuccessful attempt to stow aboard Commodore Perry's ship in a desperate effort to gain first-hand knowledge of Japan's enemies. But Yoshida Shōin's greatest significance in Japanese history was as a transmitter of Chōshū loyalist thought and activist for the imperial cause. Although his most famous students, Itō, Yamagata, and Kido Kōin, repudiated his extreme radicalism and resisted his impractical, zealous, and violent schemes for action, they subscribed to his patriotic goals and, as leaders of the Restoration, they became identified with his *shishi* ethics.[49]

The young Kawakami appears to have been drawn to that very side of Shōin's nature that the more cautious Restoration leaders mistrusted. He was captivated in 1893 by Tokutomi Sohō's depiction of Shōin as a fiery revolutionary who had been thrust into political action by the forces of social upheaval: "What he taught breathes the spirit of revolution," Tokutomi wrote, "what he preached was the method of its accomplishment."[50] The book's description of the lower samurai's role in the Restoration must have been exciting drama to an impatient young man, himself of lower samurai background, who was already vitally aware of the crucial role youth was expected to play in Japan's modern transformation. Tokutomi, the apostle of a "second revolution," summoned his young disciples with the charge of leading the reform in values and ideas that he advocated.

Another goad to Kawakami's political aspirations was the establishment of political parties and a constitutional form of government in Japan in 1889. These innovations in the political process appeared to broaden the path to power, thus encouraging Hajime in his belief that he, too, like his Chōshū forebears, might one day become a great statesman:

> The craving to become a statesman, submerged at the bottom of my heart, had been constantly stimulated by Master Shōin and, aroused by a newspaper article about the establishment of a party cabinet in Japan, that craving finally broke to the surface. My young heart was all excited by the news that party men such as Ozaki Yukio, Daitō Gitetsu, Matsuda Masahisa, and others, who were of a level unworthy to carry a silk hat, had suddenly become Cabinet ministers, and I hoped that I, too, might be able to accomplish great tasks. Literature was not a challenging enough career for a man to pursue for his entire life.[51]

The political side of Hajime's personality eventually won out: in his last year of high school, he decided to transfer from the literature to the law department. School officials at first refused to approve Hajime's new study plans. His German language teacher visited him in an effort to change his mind. "You are not cut out for law," he told him. But these blandishments merely stiffened Hajime's resolve. He notified school authorities that, if denied his request, he would drop out of his graduating class and repeat the final year of school in order to qualify for the major in law. The Kawakami family, knowing their son better than his schoolmasters did, stayed out of the picture. "Do as you wish," was Sunao's only advice. He had learned long ago that in any contest of wills, Hajime would always win. And he was right. "Once I have decided on a course," Hajime once said of himself, "I cannot be easily deflected from my goal." The school officials relented.

To prepare for senior qualifying examinations in law, students were required to take an introductory survey course in law in their senior year. With the exception of this course, the literature and law curricula were essentially the same. Not having taken the required law course, however, Hajime was forced to acquire this basic information on his own. Since there were no law texts in any of the Yamaguchi bookstores, he borrowed the single volume on law owned by the school library, cut it in half, and took turns with a friend reading each half. He passed the examination, and in September 1898 at the age of nineteen he was admitted to the politcal science department of the law faculty of Tokyo Imperial University. Looking back on this milestone in his life and on his critical decision to switch from literature to law, Kawakami remarked in his autobiography, "I had already become quite ambitious."[52]

2. crisis

Turn-of-the-century Tokyo was a vibrant, dynamic city, the largest in the nation and the center of Western learning in Japan.[1] Its main streets, with their horse-drawn trams, gas lights, and red brick buildings, were oases of modernity, where well-to-do shoppers bought American-made sewing machines, European pastries, and English Victorian clothes. Originally a sleepy fishing village, by the end of the nineteenth century Tokyo had become a major mercantile hub as well as the political heart of the country.

Not all of Tokyo had been swiftly transformed into a glamorous urban mecca. Tokyo was actually a city of striking contrasts. Stimulated by the settlement of large numbers of warriors and their attendant retinues in the Tokugawa period, the city had rapidly grown around the shogun's castle in a series of unplanned markets surrounded by clusters of wooden houses. Spurred into further growth by the hectic pace of Meiji reforms, Tokyo had continued to spread, with no apparent regard for principles of esthetic uniformity or zoning. Teahouses stood next to paddy fields and factories next to residences, creating, in addition to a variety of sights and smells, a hodge-podge of architectural styles, from the modern stone municipal buildings that housed the nation's new civil service to the flimsy shacks that served as home for most of the city's dwellers. Some former daimyo attempted to span the gap between old and new by adding English parlors to their Japanese-style manors.[2]

Differences in income were as evident as contrasts in architecture, for the rich and the poor, like the old and the new, coexisted in close, if not altogether comfortable, proximity. It was easy to distinguish the one from the other. The wealthier members of Tokyo society dressed in bright-colored silks, often of European origin; and government officials of the young nation's ministerial staff appeared at diplomatic functions wearing top hats and long-tailed black coats and carrying walking sticks. At the other extreme of Tokyo society, the city's poorer residents dressed in drab cotton or in a mixture of Japanese and Western-style clothing that reflected the general hybrid nature of city life. It was not unusual to see a young student wearing Japanese wooden clogs, baggy trousers, a T-shirt, and a garrison cap modeled on Western military uniforms.[3] While the rich shopped in fashionable stores along the Ginza, workers in the new textile factories spent as much as twenty percent of their income at pawnshops, which, if we can believe one newspaper account of the day, accepted cats, potted plants, and even cooked food as collateral.[4]

Depending upon one's personal tastes and background, Tokyo presented to the newcomer a scene of either exciting variety and modern progress or one of bewildering confusion and depressing poverty. To the nineteen year old boy from the backwoods of Yamaguchi prefecture, determined to make his mark on Meiji Japan, Tokyo was a shattering experience. With the exception of his prison term some thirty years later, Kawakami Hajime's four years at Tokyo Imperial University, together with the six years he spent in Tokyo after graduation, were probably the most miserable of his entire life.

He arrived at the capital in 1898 with great expectations and an upset stomach. It was the first time he had ever been to Tokyo or even traveled outside his home province, and he was seasick for much of the three-day boat trip. Aware that a new life dawned before him, his grandmother had stood at the gate in front of the Kawakami home when he left and with tears in her eyes had murmured, "This is the end."[5] They both sensed that his departure for Tokyo signaled the close of his youth.

When Hajime arrived in Tokyo, the capital was at the center of a brewing political storm. Only a decade earlier, the Restoration leaders had promulgated a constitution and convened the parliament, ushering in a new age of politics in Japan. But the new political system, for all its liberal trappings, was designed to maintain the oligarchs' grip over political power by blocking access to the post of prime minister for anyone not part of the charmed inner circle of Chōshū or Satsuma Restoration leaders and their protégés. After ten years of fierce squabbling between the party men in the Diet and the oligarchs, the generation born in the 1830s still vigorously resisted the notion that they should share power with others, and they had moved to strengthen nationalist ideology, centered on the emperor, in order to retain the loyalty of the emperor's subjects in the face of demands for greater representative government.

Contemporary critics leveled charges against the oligarchs for the authoritarian nature of their rule, but criticism was not confined to politics alone. There was a growing sense at least among the minority of vocal and informed dissenters that something had gone wrong with Meiji Japan. Social criticism was a new phenomenon in Japan, and although some reformers were already influenced by European socialist thought, few if any were prepared in 1900 to explain what exactly was wrong in terms of a theory of social change. Instead, different critics chose different aspects of the problem to attack and offered various solutions, depending upon their own intellectual persuasions. Tokyo appeared before Kawakami's eyes like a huge marketplace of ideas, where aspiring political leaders and determined social reformers representing the entire spectrum of political ideologies stridently hawked their wares.

Not long after arriving in Tokyo, Hajime became caught up in the intellectual fervor of the capital, running from one public speech to another and pouring over magazine articles to sample the rich diet of ideas. The city was buzzing with reformers, organizers, critics, and visionaries. Hozumi Yatsuka, an Imperial University law professor, extolled the theory of the emperor's sovereignty, and Kinoshita Naoe, a *Mainichi* journalist, denounced it. Uchimura Kanzō, a former samurai, was publishing a pamphlet on the Bible, and Tanaka Shōzō, a Diet Representative, was giving talks on Confucian revival. In any one weekend, one might find Katayama Sen organizing labor rallies, Christian women organizing relief funds, and Abe Isoo, another former samurai and a Christian pastor, organizing demonstrations against licensed prostitution and high trolley fares.

The effect of this polemical environment on young Hajime was overwhelming. Politically aware, but impressionable and naive, his mind was prepared to catch but not fully absorb the flow of words pouring from the lips of popular reformers. One imagines him sitting wide-eyed as he listened to Tanaka's vituperative rhetoric on the decline in moral standards. Kawakami was much taken by the old man who, with his flowing white beard, made a striking appearance; he always chose a front-row seat at Tanaka's lectures.[6] Anothe popular speaker of the day, though quite different from Tanaka in appearance and ideas, was Kinoshita Naoe, whom Kawakami also idolized. Kinoshita, a Christian socialist, customarily wore shabby clothes, perhaps to illustrate his nonconformist views, and his talks, given in a downtown theater, usually denounced all the shibboleths of Meiji Japan, sending shivers down Hajime's back, but also arousing in him infinite respect for the rebel journalist's courage:

> Even today I can remember the nakedness of [Kinoshita's] attack on the imperial right theory. Reared in the backwoods countryside of Yamaguchi, the original homegrounds of the Chōshū clique, I had never heard that kind of talk before, not even in a whisper. I was astounded to hear this lecture, and more than that, I was terribly shocked when the many listeners gathered there burst out in unison in great applause. I thought that if that had been Yamaguchi, the debater would have been beaten to death. That lecture opened my eyes. From this time on, the idea of democracy sprouted in my mind.[7]

Kawakami heard the leading social critics of the day challenge many of the basic premises that had guided Meiji society during the early reform period. These men were exposing the existence of pressing social as well as political problems ignored by the Meiji oligarchs. In particular, they lay bare before the Japanese public some of the unfortunate aspects of industrialization, such as low wages and cruel working conditions, which they attributed to the callous irresponsibility of government and business leaders. The very chaos of Tokyo itself bespoke a lack of rational order and moral purpose behind the nation's industrial drive. The spirit of unity championed in the early Meiji period threatened to give way to dangerous conflict and divisiveness.

Although reformers offered a bewildering array of solutions, they all agreed on the necessity for social action. One common message running through the talks of such disparate groups as the socialists, the Confucians, and the Christians was the need for a new ethical awareness and social consciousness on the part of the Japanese elite. It was a theme which, before long, Hajime would zealously adopt as his own.

The words of contemporary moralists struck a responsive chord in the young Hajime's breast because he, like so many other young Meiji intellectuals, lived his country's fate in a peculiarly personal way. But moral preachments hit home for another reason as well. Hajime's discovery of social problems coincided with the advent of emotional crises, and these two dimensions of his existence—the social and the private—both cried out for reform.

Self and Society in Meiji Japan

Almost immediately after reaching Tokyo, Hajime had run into trouble. The first of many personal problems involved his aunt and uncle, at whose home he was boarding. Kawakami's uncle, Kin'ichi, his mother's brother, was a wealthy and well-placed member of Tokyo business and government circles. A director of the Sumitomo Bank, he had been one of fifteen students to graduate in the first senior class of Tokyo Imperial University in 1879, the year Hajime was born. A Meiji equivalent of the Horatio Alger myth, Kin'ichi had served as a patron for his nephew, sending him magazines and books from Tokyo and a generous gift when he graduated from the Iwakuni school. He now agreed to pay Hajime's tuition in addition to providing room and board. Hajime's family contributed a small allowance for miscellaneous expenses.

Kin'ichi traveled much of the time on business, leaving Hajime with his wife, a woman whose views on the handling of the young were considerably less indulgent than her nephew was accustomed to. She dismissed the three or four houseboys who had been working for the couple in their elegant two-story house and required Hajime together with five or six female servants to do the menial household tasks. Although students in those days often worked as houseboys and even ricksha drivers in order to support themselves, the proud Hajime resented having to carry other people's bundles, clean the toilets, heat the bath, and eat cold leftovers in the kitchen, under the supervision of his aunt. Such indignities hardly fitted Hajime's estimation of his own self-importance. After four months, the situation had become intolerable.

Putting pride above prudence, Hajime had risked the loss of his uncle's financial support by moving to the cheapest boarding house he could find, resolving to make ends meet by working and, if necessary, getting additional money from his grandmother. When his uncle returned home, he summoned Hajime, who steeled himself for the loud scolding he fully expected to receive. When crossed, Uncle Kin'ichi's temper flared. Once again, Hajime prepared for a showdown with authority.

Kin'ichi's first words were sarcastic: "I hear you said we made you eat what cows eat," he began, referring to his nephew's criticism of the Spartan Japanese-style breakfast of pickles and rice he had been served every morning in his uncle's house. Hajime's appetite had always been finicky, but he had apparently developed a taste for Western food, such as milk and meat. Seeing the sour look on his nephew's face, however, Uncle Kin'ichi softened his tone. "How much does your room cost?" he asked and promised to continue paying his nephew's expenses. Hajime had got his way again.

With the supervision of his relatives now replaced by the company of classmates at the boarding house, Hajime had quickly assumed the lifestyle of the Imperial University student, shedding his Japanese clothing for the standard university dress, a Western jacket and a cap with his gold university pin. He woke up early, raced off to his eight o'clock classes, and after hours, browsed for books at the second-hand bookshops in the Kanda district of the city. Friends were mainly his Yamaguchi high school classmates with whom he shared lecture notes or trooped off to milk bars—the Meiji equivalent of coffee shops—for snacks, conversation, and newspaper reading.[8]

He soon discovered, however, that although he was freer, he was not happier. His mental state throughout his stay in Tokyo evidenced signs of insecurity and anxiety, a condition he himself diagnosed several years later in 1905 in a remarkably frank public confession: "For a long time I was compelled to seek for something; for this reason I was terribly anguished and insecure."[9] The physical manifestation of his nervous condition was a recurrent fever of unknown causes which subsided several years later, after he moved to the tranquil city of Kyoto.[10]

We can only speculate on the causes of his unhappiness by piecing together the regrettably small amount of information available on these formative years of Kawakami's late adolescence. Separated by many hundreds of miles from the support of his family in Iwakuni and deprived too of the personal guidance he had received from teachers in Yamaguchi, Hajime had unexpectedly found himself alone in the urban, impersonal environment of the capital city, confronted with practical and emotional problems which, at the age of twenty, he was ill equipped to handle. Harsh treatment at the hands of his relatives had aroused resentment and rebellion, but his new independence had merely created different kinds of emotional stress.

There were many dimensions to his crisis. A major problem was his inability to cope with the everyday business of living. Although he had enough money to eat meat every day ("my greatest luxury at the time") and to satisfy his sweet tooth for pastries, he was often forced to borrow money from friends, not out of real indigence, but because he was careless with money, a trait he would never lose.[11] Hajime's samurai class origins may have been responsible for his disregard of money. Once, in his first month in Tokyo, he had bargained with a book dealer over the cost of a second-hand book on economics and, having succeeded in getting the storekeeper to reduce the price, felt disgusted with himself for engaging in what he considered to be undignified behavior. Although the price was still too high, he nevertheless decided to buy the book, even though it meant returning the next day with the money.[12] The incident foreshadowed many of the elements that would later characterize his personality and his life work: his interest in economics, his aversion to the commercial ethos, and his aristocratic sense of dignity and self-importance. He would spend a whole lifetime trying to come to moral terms with the capitalist profit motive.

A more distressing area of conflict at this time revolved around Hajime's private life. He was consumed by desires for personal comfort, sexual gratification, and worldly success. Depressed by the confines of his three-mat room and envious of other students' lodgings, he moved to new living quarters and then moved again, some eight or nine times in four years at the university. He met several women during this time and possibly had intimate relations with one, but he felt, after his exposure to the moral teachings of Uchimura Kanzō and Kinoshita Naoe, that his behavior was wrong. Restless and discontented, he remembered later that he had made himself and his parents miserable.[13]

Kawakami himself analyzed his problem in Buddhist terms as selfishness and egocentrism; excessive desires for self-gratification lay at the root of his mental anguish. "The man who has property worries about being robbed; the man who has fame is anxious lest he lose his good name; the man who aspires to scholarship fears he will not advance; the man who is satisfied in love fears his mistress's affections will change. There are limits to a man's life; there are no limits to his craving."[14] Experiencing the confusion of late adolescence, without psychiatric counsellors or psychology texts to advise him, Kawakami drew on Buddhist insights to express a modern existential predicament: the absence of philosophical or religious teachings to aid the initiate in his rite of passage from youth into adulthood. The first sexual stirrings; vague anxieties over loss, failure, and death; a multiplicity of choices and options demanding decision—these were some of the complexities tormenting the young man from Chōshū. The simple *shishi* verities about sacrificing the self to the nation were no longer adequate—especially when the self kept getting in the way of the sacrifice.

The whole problem of individualism, on which Hajime's emotional drama centered, spotlights his conflict. Western ideals of equality, personal fulfillment, and political rights captivated the youth of Japan who hoped to wrest power away from the aging Meiji oligarchs. These ideals, associated with progress and enlightenment, came as fresh air into the stuffy confines of feudal life, promising liberation of the individual from his tight social nexus, with its carefully defined obligations to family and state. At the same time, few understood the full meaning of individualism—Fukuzawa Yukichi, Natsume Sōseki, and Uchimura Kanzō are rare exceptions—or how to live successfully according to its tenets. The word was almost universally associated with selfishness, which is how Kawakami came to view it, and young peoples' attempts to assert their individual will, and thereby their modernity, often took the form of hedonism, political alienation, or perhaps worst of all, in Japanese society, a love marriage, where the individual ignored his family's interests by choosing as his marriage partner someone who could satisfy his own personal desires. Such independent lifestyles were difficult to sustain, however, because few could live apart from the tightly knit Japanese society without suffering loneliness or remorse. Kawakami's emotional crisis during his university days was thus touched off in part by his inability to handle his new personal freedom.

Clearly he needed new moral and religious guidelines, for there had been nothing of a religious nature in his upbringing, Kawakami tells us in his memoirs, to prepare him for doubts about human existence. If anything, Hajime's willfulness, his ambitiousness, his striving to assert himself—the very qualities that were now causing him misery—had been encouraged by family members. His education had stressed achievement and success, but had not taught him how to cope with failure and loss. In his early life, Kawakami recalled, he had been sheltered from the realities of life and death: his paternal grandfather had died before

Hajime was born and his grandmother and parents lived to long lives. For the most part, he was raised "without tasting any religious atmosphere in the home." Family religious observances had been perfunctory: his parents had maintained a Buddhist altar for the ancestral tablets and the traditional *kamidana* for Shinto food offerings, but these had meant little to Hajime, and the occasional visits to Buddhist temples and to the graveyard where his grandfather was buried were more like outings than religious ceremonies. Even in his school days in Yamaguchi City, when he was separated from his family for the first time, he had been protected from the harsher realities of existence:

> Since [Yamaguchi City] was an isolated, small town surrounded by mountains, there was nothing in the way of social or intellectual stimulation—neither political speeches nor religious sermons—that might have caught my attention. By only listening to some common-sense Confucian lectures at school, I didn't have any particular occasion to raise questions about life problems, and I indulged myself in literature . . . spending five years in the dream of youth.[15]

Tokyo shattered that dream.

Hajime's university professors could not provide him with the spiritual guidance he needed. Part of his attraction to men like Uchimura Kanzō and Tanaka Shōzō came from simply contrasting these mavericks with the academics whom society esteemed. As professional teachers and as human beings, Hajime found, his Tokyo Imperial University mentors left much to be desired. It is even possible that his disappointment with the faculty and the curriculum contributed to his disturbed state of mind. Since Hajime eventually became an economics professor and, in his early years as a teacher, drew heavily on his own professors' lecture notes, it is worth examining the academic milieu in which he was first introduced to what became his lifelong work. A description of the intellectual climate in his university days suggests further reasons for Hajime's malaise.

Tokyo Imperial University was considered an arm of the state created for the purpose of preparing students for service to the nation. The law faculty, one of six separate colleges (the others were medicine, engineering, letters, science, and agriculture), enjoyed a particularly close relationship with the government bureaucracy, and law students were given special privileges in competition for bureaucratic positions.[16] The elite nature of the student body and the exalted status of the professors lent an air of awesomeness to the nation's first and, until 1897, only Imperial University; and students expected the very best of their teachers.

Until 1908, when a separate economics department was established within the law faculty, economics at Tokyo Imperial University was taught as part of the course offerings in the political science department. Even after 1908, many more students chose law or government as their major than economics, still a little known branch of learning, whose relevance to matters of national importance was poorly understood.[17] Economics, associated in some students' minds with banking or other commercial functions, lacked the importance and high purpose associated with government and law courses. Economics professors, moreover, tended to base their lectures on the dry repetition of basic economic theory, drawn directly from German textbooks. As a result, their lectures acquired a reputation for being dull and worthless.

Ōuchi Hyōe, who survived these uninspired lectures to become one of Japan's leading economic historians, remembered hearing his professors ask, "What is desire? What is wealth?" and then proceed to read the answers directly from their German books. Economics in Japan, he wrote, was a foreign science, consisting of formulae irrelevant to actual problems confronting Japanese society.[18] There were about seven men teaching courses in economics in 1908 when Ōuchi entered the law faculty. Many of these were the same teachers whose lectures Kawakami had heard only a few years before. Among these eminent scholars, according to Ōuchi, "Not one was interesting." Ōuchi's candid opinions even extended to Kanai En, the most famous economics professor in Japan at that time. "Kanai was great, he was famous, and yet his economics was trivial. With all my might I listened to his lectures, took notes, wrote exams and yet, I did not have the slightest idea of what [economics] was all about."[19] Matsuzaki Kuranosuke, Kanai's protégé and Kawakami's major professor, was not only boring but derelict in the performance of his duties, frequently failing to show up for his lectures.[20]

Kawakami's own criticism of his teachers was equally searing. "University professors," he wrote in 1905,

> lecture on the study of industrial policy or they translate all of Schoenberg's economics. The professor stands at the lectern, takes out his notes, and reads his translation. Then some ten students start writing and, at the end of the year, they have finished taking down the translation of a small part of the industrial policy in Schoenberg's economics text. That is all the students have . . . They go on to become university professors and publish as their own what they heard from their teachers.[21]

Bored and disappointed, Hajime doodled in his notebooks and when examinations drew near, he borrowed his friends' notes and crammed.[22]

Some of this disillusionment with university lecturers may have originated in the technical nature of the university curriculum. Coming from the liberal-arts orientation of their high school days into the professional atmosphere of the university, Japanese students even today experience the need to shift mental gears. University studies require a narrowing of intellectual interests to concentrate on specialized fields of study. Operating in the impersonal, tedious environment of the university, with its poor physical facilities and aloof teachers, Kawakami probably received very little personal guidance or intellectual stimulation.

More depressing still, the bureaucratic as well as academic world was now given over to narrow professional concerns. In fact, the whole complexion of politics in Japan had changed since Hajime's early school years. Politics was losing its glamorous appeal as government became a routinized, bureaucratic function. The administrative complexities of the bureaucratic machine placed a premium not on a person with personal initiative and boldness—the qualities associated with the shishi—but on the diligent drone. What awaited him was a petty desk job or perhaps an obscure teaching position, with little opportunity for the personal glory he dreamed of winning.

Under these circumstances, Hajime's vision of himself and his future, like his enthusiasm for Japan and its future, disappeared in clouds of doubt. The great flood of ambition that had carried him through his many years of schooling seemed suddenly to have dried up, just as he was reaching for the top rung of the ladder of success. What is more, room on that top rung was no longer guaranteed for all the many ambitious candidates for political office rising through the nation's new system of mass education. Competition for positions of prestige and influence in late Meiji society was becoming more and more intense. Social mobility constituted a new factor in Japanese life, and while the prospects for advancement were exciting, the pace of competition was grueling.

Deeply troubled by his loss of purpose and also perhaps by a fear of failure, Hajime, in his junior year at the university, entered a period of brooding introspection. Two of his friends simply gave up the struggle for success. "To put it nicely," Kawakami wrote in his memoirs about one of these friends, "Yoshikawa was a bit of a cynic," or, "to put it unkindly," he was "a straggler whom the world threw away." Yoshikawa made no effort to become famous, refused to wear the university pin that instantly identified students as potential members of the Japanese elite, and spent his time either reading or practicing Zen. As a result, he graduated at the bottom of his class, never had his period of brilliance, and ended up working in a library. Another close friend,

Okamura Shōichi, managed to obtain a bureaucratic position after graduating, but retired early, supported by rents from land he had inherited in Yamaguchi Prefecture.[23] Too proud and too poor to follow either friend's casual and even careless attitude toward the future, Kawakami spent his university days drifting, one part of him still hoping for conventional success while the other part restively searched for a more inspired goal.

A new term, agonized youth (*hammon seinen*), became the catch-all label in contemporary newspaper and magazine articles of the decade of the nineteen hundreds to describe the spiritual confusion of Kawakami's generation.[24] This was an age when Takayama Chogyū wrote, "I am a person of contradictions, a person of agony," and when Shimazaki Tōson gave free reign to the expression of such intimate and self-oriented subjects as love and loneliness.[25] The growing preoccupation of Meiji youth with the private sphere of self was an indication that they could no longer completely identify with the Japanese state. Their discovery of self, coming from their experience of independence, marked the emergence of a new generation of Japanese intellectuals with new psychological and spiritual concerns. Attracted to the modern style of lyric poetry represented by Tōson's *Young Greens* (*Wakanashū*) —his favorite literary work—Hajime too fell under the spell of romanticism.

Kawakami Hajime's childhood paralleled the growth of the Japanese nation; having reached early manhood, he suddenly became aware of the complexities of the new society and the new psyche. Under such circumstances, he was particularly sensitive to the words of contemporary moral and religious teachers who spoke of the need for individual as well as social reform and who, through their own life work and personal example, offered more appealing models than academic or government leaders. Among social critics at this time, Christian reformers made an especially deep impression on young Hajime.

Kawakami described his response to Christianity in rapturous terms: "The Bible awakened my soul for the first time . . . [through Christianity] I who had from infancy on no contact with religion, first began to approach its gates." He credited this religious awakening to Uchimura Kanzō, the Japanese Christian leader whose *Study of the Bible* (*Seisho no kenkyū*), a "small, but scholarly magazine," first appeared in 1900. Although he never met Uchimura personally, he felt that "it was entirely due to the influence of Uchimura-*sensei* that the Bible first came into my hands."[26]

The aspect of Christianity that most forcefully struck the young Hajime was the ethical teaching of the Sermon on the Mount. These words offered him the guidelines he needed to regulate his own behavior:

> When I first read [the Bible], what penetratingly moved my heart and pierced my soul with their powerful force were the words, "Whosoever shall smite thee on thy right cheek, turn to him the other also. And if any man will sue thee at law, and take away thy coat, let him have thy cloak also. And whosoever shall compel thee to go a mile, go with him twain. Give to him that asketh thee, and from him that would borrow of thee, turn not thou away."
>
> This I felt was the imperative of absolute unselfishness. My conscience bowed to it categorically. Upon reflection now it seems strange, but at that time I cried out inwardly "that is right" without asking why it should be so. I thought that the absolutely unselfish attitude revealed in the passage ought to be the true ideal of human conduct. And I felt that at the bottom of my heart I should conduct myself *literally* according to that ideal.[27]

Hajime's search for guidance in Christianity reveals the inappropriateness of the old Confucian texts and upbringing in the modern world. Confronting a flawed society and an imperfect self, he sought help in Western religious ideals, for he felt removed from his own society's traditional moral and religious teachings. The warrior's way had become obsolete in a world of factories and offices, and the emphasis on self-negation hardly seemed meaningful in a society that now encouraged profit making and personal ambition. Coming of age at the turn of the century, when the Meiji goal of an independent nation was close to fruition, Hajime abruptly faced the need for a more universal ethic to guide himself and the nation in the industrial age.

Yet, we can see a residue of traditional idealism in his interpretation of the Sermon on the Mount. The meek, almost submissive role enjoined by Christ's words was interpreted by Hajime as a clear call to selfless action, resonant with the ethos of the shishi and compatible, too, with his own psychological tendency to adhere stubbornly to a chosen course of action. It is therefore possible that young Hajime's attempt after this time to become a moral spokesman for his age was modeled on the samurai's traditional role as moral exemplar. We may even surmise that Christian teachings appealed to him precisely because these articulated an ethics of self-negation and self-sacrifice that echoed the traditional way of the warrior. His education in the shishi ethos of self-sacrifice paved the way for his receptivity to Christian ethical teachings.[28]

The universality of Christian ethical doctrines, however, introduced potential conflict with previous standards of conduct. As Hajime soon discovered, his new standard of behavior clashed with the priority given to the family in Confucian teachings. Moreover, the Sermon on the Mount, as Hajime interpreted it, seemed to demand the total annihilation of all desires, an ascetic ideal that went well beyond Confucian norms of proper behavior without necessarily promising any worldly rewards for the effort. "In my mind," Kawakami recalled, "there was an anxiety that I might destroy myself if I adhered to such an ideal. Thus, for the first time a doubt was implanted in me about how I should regulate my life."[29]

Despite his initial hesitancy, Hajime plunged into his new calling with a characteristic thorough-going zeal that far exceeded anything expected of him by the leading reformers of the day and that suggested he was intent upon stamping out selfishness in himself as well as in the rest of society. The Ashio Copper Mine incident provided Hajime, then in his senior year, with an occasion to translate his new ideals into action. Three hundred thousand farmers were left destitute after poisonous wastes from the Ashio mine, dumped into nearby rivers, had polluted waters used to irrigate the populous rice-growing region of the Kantō Plain. The ethical failure of government and business leaders was demonstrated by their refusal to accept responsibility for the misfortune or come to the aid of the ruined farmers. When largely through Tanaka Shōzō's efforts the disaster was disclosed before the public, a large-scale relief program was organized, uniting Christians, socialists, and Tokyo Imperial University students in a spontaneous outpouring of humanitarian goodwill.[30]

Driven by Tanaka's fiery speeches and his own awakened social consciousness, Hajime wrote an article on the affair for his home town newspaper, the *Bōchō shimbun*. The article reveals the affinity Kawakami found between the Sermon on the Mount and shishi ethics, for in it he evoked time-honored samurai values and the teachings of Yoshida Shōin in the interests of a new vocation, that of social reformism:

> Since it is the land of Bōchō that gave birth to the shishi beginning with Yoshida Shōin, I cannot believe that all of us are spineless. Let's spur each other on to action . . . What we need in this time of great emergency is men of extreme sincerity who will throw away their lives, dying for their country . . . enlightened men who cherish duty . . . What sort of human beings do you prefer? The politician who uses public funds for his own profit and desires, or the man who sacrifices the ruler's profit? As a school principal or teacher, do you like someone who fawns all over you in order to protect his position? Do you like the man who sacrifices the people for himself or the man who dies for you? If you would have others die for you,

then you must die for them ... The sacrifice of self—sincerity, devotion, in all times, in all tasks—is necessary, but it is especially necessary in the present world, a world where conditions are so terrible ... In the words of Yoshida Shōin, "A man who is extremely sincere and yet who does not act, is not yet a man."[31]

A few months later, in the fall of 1902, Hajime attended a fundraising benefit sponsored by the Christian Ladies Moral Reform Society, which had formed a Women's Ashio Copper Mine Poison Relief Organization. Kinoshita Naoe, Tanaka Shōzō, and other noted orators and polemicists of the day urged the audience to help the impoverished farmers survive the coming winter.

What followed became legend in the life of Kawakami Hajime. Filled with "wholehearted sympathy and outrage" and, as he related it, with the words of the Sermon on the Mount still ringing in his ears, Hajime decided without hesitation to donate everything he owned to the victims of the disaster. Removing his overcoat, *haori,* and muffler, he handed these to one of the women in charge at the door and, returning to his lodgings, he gathered together the rest of his clothing, except for what he was wearing, and had a rikisha driver take the bundle to the office of the relief organization on the following day. Such wholehearted sacrifice on the part of the anonymous student was viewed as sheer insanity in the eyes of the Christian Moral Reform Ladies. "I can't just keep these," their leader protested, shaking her head as she sifted through the articles of clothing that had been deposited in her office. "They say he's a bit touched in the head."[32]

With the help of Kinoshita, the fund raisers traced Hajime back to his former lodgings, where the landlord informed them that the young man had "a fine, upstanding uncle ... and so he's probably not in any trouble."[33] The story about the young man's generosity appeared in the newspapers the following day, yet when his family and friends heard about it, they were far from congratulatory. "I had purchased the derision of my friends and the rather intense anger of my mother."[34] His mother, it seems, had sewn some of the donated garments, and his younger brother had bought the overcoat for Hajime out of his own salary. Hajime learned from this incident that, if he were to follow Christ's teaching literally, he could not be filial to his parents, and so he decided he had to "turn away from the words of the Sermon on the Mount for a while" as he prepared, at graduation time, to make his way in the world.

We do not know how well Hajime's family understood the reasons for their young heir's behavior.[35] They were already familiar with his headstrong personality and his ready emotionalism. We can only assume they had noticed symptoms of uneasiness in this promising young man, such as his uncertainty about the future, for Hajime's sense of distress, so evident in his own writings at the time as well as in his recollections in later years, lay at the basis of his preoccupation with ethical behavior and social reform, confusing his practical ambition and dictating his method of solution to these problems by a combination of Christian humanism and samurai zeal.

We can be certain, however, that Hajime himself recognized the changes that had come over him as he approached graduation day. Looking back over his life some four decades later, he pinpointed this moment in time when he first confronted the awesome demands of the Sermon on the Mount as the start of the spiritual quest that influenced his entire political future: "From here on began the anguish of my soul [*kokoro*]; it is correct to say that this was the beginning of the history of my soul."[36]

3. meiji dropout

Like Hamlet and other modern men, Kawakami Hajime's life was forever complicated by forks along the road demanding critical choices but providing few signposts. One of his most difficult decisions concerned selecting a career, and one of the easiest, taking a wife. These two major events in his life occurred almost simultaneously in 1902 when, several months after graduating, Kawakami was summoned home to Iwakuni for an arranged marriage to Otsuka Hide. Shortly after the wedding, he brought his bride back to Tokyo, where they moved into a new house and he looked for a job.

The couple was well matched. Hide's home was also in Iwakuni and her father, like her father-in-law, had been a lower-ranking samurai in the service of the Kikkawa house, had fought in the Restoration wars, and after sickness forced his early retirement from the Home Ministry in Tokyo, had returned to Iwakuni to work as a steward for the Kikkawa family for the remaining years of his life.[1]

There were eight children in the Ōtsuka family. One of them, Hide's eldest brother, introduced Kawakami to her, saying perceptively, "Hajime is like a wild horse. You have to use the reins often."[2] She tried, but did not often succeed, in balancing her husband's impetuosity with her own practical-mindedness. Hide once observed that Hajime was often tricked or deceived by other people. He defended himself by saying he and his wife approached people from opposite vantage points. "When I first meet a man, I give him ten points; that is, I am at first satisfied with any man I meet. Men are all good, I think, and so I am often disappointed later on." Hide was just the opposite: she gave zeroes first, until the person proved himself.[3]

For all her practical qualities, however, Hide was unable to calm her young husband's troubled spirits; indeed, their marriage, rather than creating the stable life he needed, only exacerbated the conflicts that raged within him. His naive confidence in the government's domestic leadership shattered, his course for maturity uncertain, Hajime faced the future driven by a burning idealism which now competed with the practical necessity of supporting himself and his wife.

Hajime's immediate problem was finding a job. Despite personal confusion and lack of interest in his course work, he had managed to graduate sixth in his class, and while his class standing was not high enough to qualify him for a silver watch awarded each year by the emperor, his academic performance was certainly far from unimpressive.[4] Nevertheless, for reasons that remain not altogether clear, his initial efforts to find suitable employment failed. He may have even been resisting the natural career route of Tokyo Imperial law graduates, hoping to avoid bureaucratic service in favor of what he considered loftier work. Around the time of his marriage in September 1902 he wrote a letter to one of the Japanese socialists, Katayama Sen, voicing his concern over social problems and expressing his disturbed state of mind regarding his future life work. This document, one of the few extant examples of Kawakami's writings before 1905, is quoted below almost in its entirety to provide a rare glimpse into the young student's intellectual interests and tender emotions, as mediated by his graceful epistolary style:

> I am a mere student who has not yet had the honor of knowing you, sir. Please pardon my presumptuousness in precipitously writing to you. I pray for your forbearance.
> Sir, there is a temple in front of the place where I live. In the temple are many graves . . . It is a lonely place even during the day except for monks whom I occasionally see. In the temple grounds are men who make their living by catching [wild] dogs and killing them. They chase dogs into the graveyard and then club them to death.
> It is already around midnight. I have just finished Malthus' *Essay on Population* which I happened to pick up. I am much impressed by it. Just as I finished reading it, I heard the yelps of dogs and the sound of someone clubbing them. All the dogs in the neighborhood were alarmed and began to howl. The sound of the beating increased and the dogs' cries gradually disappeared. Those whose job it is to kill dogs have earned several coins. The sad cries of dogs about to die were terrible and make my skin cold. However, it would be a mistake to hate those whose job consists in killing dogs simply because I have a sympathy for dogs. Allow me to weep for the imperfection of present society which has produced some fellows who cannot maintain their livelihood without being forced to engage in this miserable work.
> Is there anyone who wants to terminate life for their own personal gain? Ah, the present social organization, which leads some of our brothers to bear the unbearable, cannot be worthy of my ideals. If Malthus' words are true, does this mean that population increases every year and poverty adds to the misery forever? But, the Creator's plans are far-reaching and masterly, I think.
> I do not take a gloomy view of the future of the world, but at the same time, I cannot be satisfied with the present society. Should I

not make it my responsibility to plan the improvement and progress of present society? It is a well-known fact that you have dedicated yourself since your early days to improving society and many people have admired you. I have an acute urge to follow you and your followers by venturing to contribute my feeble efforts.

As for socialism, I have just started to study it. I do not know whether or not it is feasible or beneficial. However, there is no doubt that I have certain doubts about the present social organization and that I have great sympathy for some of our pitiful brothers. This evening, I felt full of deep emotions and so I added some oil to the lamp and dared to write to you.[5]

Finding the appropriate vehicle for his social reform urges taxed the resourcefulness of the young bridegroom. Journalism might have provided one outlet for Hajime's reformist energies, since newspapermen were among the most vociferous social critics. Hajime actually went to see the Christian socialist Kinoshita Naoe, hoping for a job with the *Mainichi* newspaper and volunteering to work for a minimal salary, but before Kinoshita, a writer for the paper, could help him, Hajime's uncle stepped in to squelch the plan, and Hajime apparently offered little resistance. An inquiry about a job with the Mitsui Bank was never answered nor did other prospects materialize. Perhaps because Hajime was torn between his new aspirations as social reformer and his family's wishes for a more conventional career and perhaps, too, because he himself was not certain which route to follow, in the end he allowed the dictates of filial obligation and social respectability to win out. Through the good offices of one of his teachers, Matsuzaki Kuranosuke, a major figure in the world of economics, he found himself before long teaching agricultural economics in Tokyo Imperial University's faculty of agriculture.[6]

In the next few years, Kawakami struggled to reconcile his new career with the voice of his conscience. Two demons tore at him: a driving ambition and a nagging idealism. Whereas in his childhood both had been joined in the goal of government service, now the very nature of his work established a conflict between the two. Hajime still wanted desperately to achieve worldly distinction, but he wanted to succeed through noble pursuits, and just as he had early in his life rejected literature for not being worthy of him—literature lacked a high enough moral purpose—so he subjected the study of economics to close moral scrutiny and found it too was lacking.

Economics as a Moral Vocation

The academic study of economics was in its early stages in Japan when Kawakami was a university student, and not many of the nation's aspiring young candidates for political office understood what exactly distinguished economics from other fields of study. The Japanese word for economics, *keizai,* was derived from the Confucian term for statecraft, *keikoku saimin* (lit., rule the country and help the people). Among Tokyo Imperial University professors economics connoted political economy or public policy, rather than a purely value-free, descriptive science of economic laws. Their treatment of economics, closely following the German approach, ranged broadly over areas that today would fall more properly in the category of public administration and betrayed a definite nationalist bias, best epitomized by Kanai En, the leader in the world of economics, who named his eight children in consecutive order, Kei, Koku, Zai, Min, Sei, Gi, Shō, and Ri, to spell out the motto, "Rule the country, help the people, justice and victory."[7]

At first, Hajime tried to glamorize his new career by linking it with the goals of statecraft. "The warrior, grasping his sword, startles the world," he wrote in 1905, in one of his first published works, "but the scholar, on the contrary, can subjugate the realm with his pen."[8] Hajime thus resolved to "make a flying leap into the academic world as a patriotic statesman."[9] Although economic study was still the handmaiden to actual bureaucratic service, he recognized even at this early date the potentiality of this new field of study as an arm of government service. Precisely because the discipline was as yet little understood, he had a chance to do pioneering work that would contribute to national development while at the same time forging his own name as one of the great men of the period: "I must make my name go down in history by discovering a great principle in economics that has not been discovered by my predecessors."[10]

Such rationalizations about the career which, in a sense, had chosen him failed to allay his doubts about its moral validity. It would take Kawakami many more years before he would discover a way to use economics in the service of humanist goals; in the years between 1902 and 1905, he was learning that his professional studies inhibited his reformist impulses. His lectures and scholarly writings repeated the material presented by his professors at Tokyo Imperial University; these were the very lectures that, in his university days, he had found dull, imitative, and useless in addressing social problems. Eventually in 1905 he railed against such teaching methods, concurring at least temporarily with the words of Henry George that, "We can not safely leave politic [sic] to the politicians or political economy to college professors. The people themselves must think, because the people alone can act."[11]

Kawakami's doubts about a career in economics thus stemmed in part from the disappointing performance of the most respected Japanese economics scholars of the day in tackling the needs of industrial workers and to their connivance in persecuting the one group most actively championing the workers' cause. Suppression of the socialists' freedom of speech and press turned him angrily against the government and those who supported the government: "The *National Academic Society* magazine (*Kokka gakkai zasshi*) preaches academic freedom." he wrote in 1905, using an assumed name to protect his anonymity, "but if freedom of speech is necessary for scholars, it is also necessary for the people."[12]

A more fundamental reason for Kawakami's uneasiness about economics as a vocation was related to the changing nature of economic study itself. We can obtain some idea of his thoughts on the matter by jumping ahead to 1905, when the fruits of his labors over this three-year period after graduation were realized in a harvest of publications. As he familiarized himself with a variety of Western sources, he began to realize that economics was becoming a narrow science that emphasized the material aspects of human existence to the exclusion of all other factors. "Economics studies only one kind of social value or social utility. This is because it is impossible to explain all kinds of social value or social utility. The facets of man's activity are varied and complex."[13]

The one-sided nature of economics, based as it was on the description of how men satisfied their material wants, neglected those fundamental questions of morality that preoccupied Hajime's mind. The growing trend in Western thought to concentrate on economic life as the basis of all human activity was clearly spelled out in a new work entitled *The Economic Interpretation of History*, published in 1901 by Edwin R. A. Seligman, a professor of political economy at Columbia University. In the introduction, Seligman called attention to a new "tendency in recent thought" to demonstrate the "essential interrelation of politics, ethics, and economics." Calling this new doctrine the "economic interpretation of history," Seligman predicted that it would "spread to the uttermost limits of scientific thought." This new system of thought treated the reasons for social development and human progress in terms of economic causes: "To economic causes . . . must be traced in the last instance those transformations in the structure of society which themselves condition the relations of social classes and the various manifestations of social life."[14] Tracing the history of the idea of physical causation back to the Greek philosophers, Seligman eventually arrived at Marx's contribution to the notion that "social life at any one time is the result of an economic evolution."[15]

Seligman's lucid account of recent intellectual currents, written in a simple prose style with a genial, persuasive tone and literary flourish that bespoke its author's greater debt to the humanities than to science, captured Kawakami's attention and became his first published translation. The book also represented his first brush with Marx, from whom he recoiled with some distaste. Although he confessed in his autobiography that he translated the book without fully understanding the meaning of historical materialism, judging from his discussion of the work in 1905, he evidently was troubled by some of its implications.[16]

The world view represented by the economic determinists treated man as merely an economic animal. It was a view in which men, as individuals or as members of economic classes or nations, competed with other men, classes, and nations for the limited material goods available in the world. Economics consisted in describing the laws of this competition and showing the relation between material resources and men's desires. Progress was based on the increase in material goods, and this increase was, in turn, based on man's drive to satisfy more and more of his desires. Social development depended ultimately on whetting men's appetites.[17]

Such a view of human society and human history jarred Kawakami's moral sensitivities. It left no room for the kind of moral preachments that he had associated with statecraft and hoped to bring to the study of economics. Man's work has been called a "uniquely important boundary between self-process and social vision." As Robert Jay Lifton describes it, adult work is "conducted under the guiding principles of the culture's assumptions about transcendence": it is "tied in with a larger spiritual principle."[18] What was the "larger spiritual principle" embodied in modern economics? As far as Kawakami could see, there was none. "Man is a selfish animal," he wrote. "We are generous with other men's things and unnecessarily sparing with our own. The present system of private property utilizes this weak point."[19]

Given the overriding importance of economic behavior in human history, Kawakami recognized that it was natural for economists and historians to emphasize material factors in explaining the evolution of human society. Still, he faulted the economic determinists for failing to take into account other than purely material factors in explaining men's motivations. To view history solely in terms of economic behavior was to neglect altogether the spiritual side of human existence:

> Just as there is a variety of kinds of human activity and desires, there are various ways to explain history. There is not only the economic explanation, but there are, in addition, the ethical, esthetic, political, legal, linguistic, religious, and technological explanations. This is why each scholar can view the happenings of the past from his own vantage point.[20]

It was true, Kawakami added, that economic explanations tended to carry more weight in the present world of limited natural resources. At the same time, he expressed faith in the future when, owing to improved technology, men would acquire enough material goods to satisfy their desires without hostile competition:

> When we reach a better world, where science controls population, where men do not compete, where goods are available, and so on, then economic conditions will lose their importance. Until that time, we must place more weight on economic conditions than on any others ... The economic explanation is relative, not absolute. Its theory best suits the past; in the future it will gradually lose its importance. It does not explain all progress; it just emphasizes the fact that economic circumstances seem most related to the rise of nations and peoples.[21]

Kawakami's biographers tend to brand his earliest writings as unoriginal and therefore unimportant, influenced as these were by the German school of nationalist economics. Kawakami, too, in his later years, derogated his initial efforts as an economist. What is often overlooked, however, is the fact that Kawakami's lifelong intellectual concerns were prefigured in these initial publications. His effort to retain normative considerations in interpreting the past as well as guiding the future became the basic theme of all his scholarship.

Kawakami's keen interest in the reasons for historical change and his agreement that technological innovation and economic factors greatly influenced the course of history represented views generally held by Meiji intellectuals and government leaders. Their country had begun its massive reform movement under the guidance of samurai who had become acutely aware of Japan's technological backwardness. Economic development, utilizing the latest advances in science and technology, was the goal of Meiji reformism. At the same time, Kawakami's equally strong conviction about the importance of the spiritual as opposed to material side of human life marked his coming of age in a period when, it seemed, material technology and materialist values dominated men's consciousness. Having survived the threat to its existence as an independent nation, Japan was now being eaten from within: immorality—selfishness—was rampant. The cure had proven as bad as the illness. Whereas technology was the answer to Japan's material backwardness in the decades of the eighteen sixties and seventies, spiritual reform was necessary in Kawakami's time to temper the ill effects of Meiji economic reforms.

Kawakami's early writing thus displays a tension that characterized most of his later thought: receptivity to economic determinist interpretations of history to the extent that these explained the progress of society but resistance to such explanations insofar as these excluded from consideration questions of morality. In philosophic terms, Kawakami was a dualist, reluctant to accept the monistic world view advanced by Social Darwinists, Marxists, German economists, or any scholar, for that matter, who neglected the role of spiritual forces in explaining the course of human history: "It is clear that we cannot fully explain history by studying only economic relations, since human history has been influenced by spiritual forces ... It is equally clear that the basic relationship between social organization and social classes is manipulated by economic circumstances."[22] Reconciling the material and spiritual values of human life was eventually the task which Kawakami Hajime set for himself as scholar-statesman. But first he had to work out this equation in his own personal life.

Kawakami Hajime in 1902 was still determined to make a name for himself if not in government then in the academic world, and he threw himself into his studies, burying his doubts in an avalanche of hard work. His immediate sights were set on winning a government scholarship for study abroad, the sine qua non for advancement, but his efforts to familiarize himself with economics were thwarted by personal as well as philosophical problems.

Through Matsuzaki he received offers to publish articles on economics and enough outside jobs to meet his expenses, but, obligated to his mentor, he soon became the older man's lackey—running errands for him, doing research for him, helping him organize his private library. Kawakami bitterly recalled how, "From breakfast time on I was piddling with books."[23] Although Matsuzaki took care of him he never left him alone. To make matters worse, Kawakami's younger brother and two of his cousins came to live with Hajime and Hide shortly after their own son was born, and the hubbub in the house made study impossible. With his chances for achieving worldly distinction—his period of brilliance—growing dimmer every day, Kawakami, never one noted for patience or quiet resignation, temporarily dissolved his family in 1905, sent Hide, who was pregnant again, to live with his parents, moved to different lodgings, and resolved to dedicate himself anew to his economic studies.[24]

In the burst of energy and gift of quiet that followed his decision to return to bachelor living, Hajime's scholarly output dramatically increased. His translation of Seligman's *Economic Interpretation of History* (*Shinshikan*), which introduced the concept of economic determinism into Japan, was well received by the academic community, though an otherwise favorable review by Yoshino Sakuzō pointed out that Kawakami had inaccurately translated the "dialectic" as "the law of progress."[25] Nevertheless, the book stirred some young people's interest in the little-known subject of economics: as one future economist recalled, "I realized somehow that if you do economics, you can understand history."[26]

During the time that Hajime was living alone he also served as an editor for the *National Academic Society* magazine (*Kokka gakkai zasshi*), a position secured for him by Matsuzaki, and completed three books: *Fundamental Principles of Political Economics (Keizaigaku jō no kompon kannen), On the Reverence for Agriculture in Japan (Nihon sonnō ron),* and *Theories of Political Economics (Keizaigaku genron).* The first one, only forty-three pages long, was published at his own expense.[27] A good part of this little book was devoted to analyzing the nature of economic acts and to distinguishing between economic acts and other forms of behavior. These niggling theoretical distinctions owed much to German scholarship but also reflected Kawakami's own nagging conscience. He seemed disturbed by the failure of economists to recognize the varying motives underlying economic acts, and he felt a need to rescue at least some kinds of economic behavior from their origins in "desires," a word he equated with selfishness.

Kawakami argued that "economic desires do not necessarily emerge from selfish motives," because in certain cases, such as philanthropy, a man's desire to make money in order to do good works was an economic act motivated by a charitable heart. Similarly, not all acts motivated by economic desire were economic acts. Using an example close to home, he explained that the desires which controlled the economic acts of scholars were actually academic desires, whereas the desires governing the economic acts of merchants were economic desires. The desire for education was not an economic act, nor was the acquisition of education. But, when the scholar bought a book or sought a job, he was engaged in an economic act, exactly like a merchant buying goods. Motivation was thus the key to defining the nature of an individual's behavior: "In order to determine whether a certain act is or is not an economic act, you must look at the immediate desire which gave birth to that act."[28]

It is difficult to see the theoretical significance for the science of economics of these nit-picking definitions, but they are interesting, nevertheless, for the insights they yield into Kawakami's mental state at the time he wrote them. While discussing theories of economic behavior based on the motivation of the actor, he became all the more sensitive to his own motives for writing about such theories. The more he dedicated himself to his economic studies, the more he agonized over the value of his work and his reasons for doing it:

I came to the realization that the path of economic research which I had chosen in order to raise my status was, in the end, nothing but ... the search for my own fame and profit, and the very opposite of absolute unselfishness. Once again doubts concerning human life—and the anguish because they were still unsettled—pushed to the foreground of my consciousness. Although I had gone through the trouble of devoting all my energies to economics, ... I was seized by doubts as to whether I shouldn't give it all up.[29]

Ambition and virtue, so casually joined in the school curriculum of Yamaguchi Prefecture, were now on contending sides of Kawakami's value calculus. Was not the importance placed on self-assertion, fame, and ambition merely symptomatic of the more basic problem of egoism? In encouraging young people to rise in the world and make a name for themselves, Meiji leaders had encouraged an epidemic of moral insensitivity. While all of Japan was rushing along a suicidal course, he, as an economist, was not only preaching a false faith, his nation's leaders were following it. Before curing Japan, he first had to cure himself.

In the course of his studies, Kawakami had read about an Oxford University economics professor and English social reformer named Arnold Toynbee (1852–83), who had moved into a slum district in East London to help educate impoverished residents in the area. A pioneer university settlement house, Toynbee Hall, was named after him to commemorate his dedication to welfare work. Once again, Kawakami was moved by the personal sacrifices of a man of seemingly selfless motives. Toynbee, uncle and namesake of the famous English historian, "failed as a scholar," Kawakami observed, "but as a human being he succeeded. After all, there is no work so great and praiseworthy as that of leaving behind a pure, beautiful life history."[30]

Toynbee's life now served as an inspiring example of Kawakami's newly evolving standards. He saw the choice confronting him as one between gratification of personal needs and compliance with duty:

Should I continue to be an instructor in economics or should I, like Toynbee, resign my position and engage in the education of the poor? Should I chase after fame or wait upon benevolence? Should I seek profit or obey duty? I had to choose one or the other. A countless number of times I sought to delay the solution of this problem. Yet ... this was something demanding resolution and could not be delayed, even for one moment. Nevertheless, leaving one's teaching position is no easy matter, and so I hesitated and vacillated, one way and then the other.[31]

These questions about his life work pushed Kawakami at the end of 1905 to the brink of emotional collapse. His accomplishments over the previous years had been achieved at great cost. Separated from his family, tired, harried, and lonely, he had condemned himself to working until the early hours of the morning in order to complete his studies for publication. He began to suspect that to achieve his ambitious ends would require "a number of years of weeping in misery, and this I cannot bear."[32] A more immediate success might have lessened his strain or made it all worthwhile; but caught between the desire for fame and the costs of achieving it, he felt the crisis that had been simmering since his university days reach a boiling point.

The crisis occurred as Kawakami was in the middle of writing a series of articles on socialism for the *Yomiuri* newspaper. He had made elaborate efforts to conceal his identity, for fear that the opinions he expressed would not find favor with the government and he would jeopardize his chances of winning a Ministry of Education fellowship.[33] Nevertheless, the clever polemics and well-turned sentences in Kawakami's *Critique of Socialism* (*Shakaishugi hyōron*) quickly captured the attention of the *Yomiuri* readership, comprised largely of students and the Tokyo intellectual elite, and his unfamiliar pen name, Senzan Bansuirō, instantly sparked their curiosity.[34]

Kawakami's critique of the young Japanese socialist movement reveals the impasse he had reached in his own thinking about the ills of Meiji society and his own role as society's custodian. In his vitriolic attacks on contemporary social and political leaders, the mysterious Senzan left no quarter unscathed.

He began by condemning the establishment. The government was unfairly persecuting the young socialist movement, and Tokyo Imperial University professors were merely lackeys of the state, who criticized others in order to make themselves look good.[35] He then criticized reform leaders for the inadequacy of their efforts. Japanese socialists such as Kōtoku Shūsui, he argued, failed to appreciate the need for moral as well as institutional reform; whereas Japanese Christians like Uchimura Kanzō went to the other extreme, occasionally "spew[ing] forth Biblical sayings," but ignoring altogether the importance of concrete reform measures.[36]

Christian socialism might conceivably have provided the two emphases which Kawakami thought were needed—the spiritual revolution of the individual and the material improvement of society—but the Christian socialists, Kawakami charged, as he neared the close of his series of articles, were guilty of an impure and incomplete faith. Although they claimed to believe in God's love, their efforts to improve the social system showed their dissatisfaction with God's love.[37]

This criticism of the Christian socialists owed much to Leo Tolstoy, whose celebrated letter to Abe Isoo had caused a great stir among young intellectuals when it was published in the socialists' *Commoner* newspaper (*Heimin shimbun*) a year earlier in 1904. Tolstoy, who was actively corresponding with other Asian reformers like Gandhi, had written that social harmony came about only through the religious and moral perfection of men as individuals. The socialist movement, Tolstoy claimed, was meddling with the single law of truth, which was simply to love all men.

How did advocating the single law of truth differ from "spewing forth Biblical sayings"? Kawakami's critique had by its sophistry demolished all roads to reform. In his demands for answers satisfactory to all his quests, he had once again pushed himself into a dead end. Satisfactory answers were not forthcoming, because it was no longer possible for anyone to grasp the totality of the changes that were overtaking Meiji society and authoritatively defend any one right method of reform. Yet, this is precisely what Kawakami needed—a single answer that would direct all the choices he faced. He needed a new view of man and society that could demonstrate how spiritual and material progress were linked, so that he could comprehend the changes in human values and technology that were overtaking himself and Japan. It was characteristic of his generation, however, to seek answers to social problems in religious leaders rather than in theoretical constructs.

Kawakami's search for religious guidance eventually compelled him to abandon altogether his academic work, in quest of a moral exemplar who combined Christian humanism with practical activism, someone as totally dedicated as Yoshida Shōin. Ebina Danjō's talks on Christianity at first captured his interest, but observing the aristocratic deportment of this Japanese Christian and former samurai, on whose kimono the family crest was elegantly printed, Hajime cynically doubted the sincerity of Ebina's commitment to Christian charity.[38] Tolstoy was a more convincing model, because he had actually renounced his aristocratic lifestyle to live among his serfs. Wandering through a second-hand bookstore one day, Hajime came across Tolstoy's *My Religion* and reading it, felt "all through the thirty chapters . . . as though I had been struck by electricity."[39]

Like Kawakami, Tolstoy too had been both inspired and troubled by the Sermon on the Mount, in which "there was demanded a too impossible renunciation of everything, which destroyed life itself . . . and so the renunciation of everything, I thought, could not be a preemptory condition of salvation." Tolstoy suggested at the end of his book that the clear and simple meaning of Christ's message was "do not resist evil" and love mankind.[40]

With Tolstoy's life example as final encouragement, Hajime determined to start a new life. In December 1905 he resigned from his position as a college lecturer in order to dedicate himself to teaching the poor. By continuing to publish his series of articles on socialism in the *Yomiuri* newspaper, he retained a small income and, even more important to him, a forum for publicly expressing his ideas, though he still did not know how to give concrete shape to his vision.

Several days after reading *My Religion,* Kawakami uncovered a religious tract called *Selfless Love (Muga ai)* published by Itō Shōshin, founder of a new communal sect known as the Garden of Selflessness *(Muga En).* Thinking he had found in Itō a practitioner of Christian humanist ideals, Kawakami immediately contacted him, gaining permission for an audience three days later, on December 4, 1905.[41]

The Garden of Selflessness

Itō Shōshin, only three years older than Kawakami, had himself undergone a religious crisis less than one year before. Trained for the priesthood in the True Pure Land sect of Buddhism, he had become disillusioned with the emphasis on success and achievement that had penetrated even the gates of Buddhism. His father's illness had served to crystallize his own thoughts on the meaning of life, leading him to reject all organized religion in favor of a general humanism based on love for "natural mankind." Leaving the True Pure Land sect, he had vowed to set an example of selflessness for others to follow. The communal organization he established in Sugamo, a small town on the outskirts of Tokyo, in June 1905 attracted at the end of the first year of its existence about one hundred persons from all parts of the country and all walks of life who welcomed the opportunity for religious retreat that the Garden of Selflessness provided. Itō's teachings, an amalgam of Buddhism, Christianity, and Tolstoyan humanism—Itō had also read Tolstoy's *My Religion—*provided the common philosophical basis for the group members, who gathered daily in the large meeting hall for spiritual meditation.[42]

Two sessions with Itō Shōshin gave the overwrought Hajime the courage to pursue his religious quest. Itō agreed with him that economics was an unworthy pursuit (he called it trivial work), but he thought no more highly of the ideal Kawakami had expressed of becoming a "repairman of society." Itō advised him instead to put all his efforts behind teaching others his philosophy of universal love.[43]

The willing disciple accepted this advice, and with encouragement from Itō, summoned the strength at last to renounce all worldly vocations. Readers of his *Critique of Socialism* were startled to find, in place of the expected thirty-sixth installment, a public letter revealing the writer's true identity and his decision to discontinue writing in order to proselytize the "truth of selfless love."

The tone of the letter reveals that Kawakami had made his decision in a highly unstable emotional state. Where the articles on socialism themselves had been academic, lucid, clever, in places biting, but always controlled, the letter seemed to be a release of everything he had suppressed in order to write them. He summarized his life up until that moment, mentioning such intimate details as his unhappy life in his uncle's house, his relations with women while at the university, and his doubts about his career as an economist. He painted a painfully frank picture of himself, confessing his ambitiousness, his loneliness, and his love for his wife. The entire piece must surely have been written in a single outpouring of pent-up emotion.[44]

The message contained in this public confession and also in an accompanying tract entitled *The Truth of Selfless Love (Muga-ai no shinri)* preached the folly of striving for worldly success, the futility of pursuing fame, and the wisdom of selfless love. Only by becoming desireless do men gain happiness and peace of mind, Hajime told his readers. They should "entrust their fate entirely to others and, at the same time, devote all their strength to loving others."[45]

These religious sentiments derived from an eclectic blending of Buddhist and Christian doctrines, but one thing was clear: Hajime sought to bury the ambitious goal of rising in the world that had driven him since childhood, while retaining the idealism that had also directed his vision. Thus, he encouraged his readers not to worry about their parents' amibitious expectations for them. His readers might feel that "because my father wants me to be a success in life, and to become rich, giving peace and joy to my parents is my single pleasure, and were I now to put into practice selfless love, my parents will surely be disappointed." Hajime countered such arguments on behalf of Confucian filial obligations by pointing out that worldly success would not give peace of mind to one's parents; what would comfort them was knowing that their child had faith.[46]

With this final assurance to his readers and also, perhaps, to himself, Hajime bid a temporary farewell to the public. Even after this bold announcement of religious conversion, however, he hesitated to join the Garden of Selflessness, because he vaguely sensed the difference between Itō's moral relativism and his own preoccupation with moral absolutes. The decision confronting him took on enormous significance, especially since he had already given up his source of income and, presumably, his opportunity for worldly success, and he was now completely unmoored and alone. It was, in fact, this very question of whether to join the communal sect or independently spread the doctrine of selfless love that precipitated the strange experience that was to influence his entire life.

The incident occurred late in the evening of December 8, after he had returned from a second visit with Itō. Alone in his room, he mulled over Itō's advice about devoting all his energies to the love of others. The hour grew late and still the question kept sleep away. How could he lead a perfectly unselfish life? He would have to give up all those activities which benefited himself, including sleep, for sleep served to rest his body and consequently was a selfish activity. To deny his body sleep, however, would lead quickly to death. Following to its limits the demands of absolute unselfishness, Kawakami once again reached the same impasse as Tolstoy—the "all too impossible renunciation of everything," including ultimately life itself.

Frail in health, he knew that, without food and rest, he would not survive for long. Yet, all of a sudden, "I firmly resolved to rush into such a way of life. At that moment, I actually determined to fling away my life. It's not that I was thinking about death, but that I willed death." This moment of revelation was followed by violent pain in his head and a contraction of his muscles. For days afterward, his body was numb; he felt no physical sensation, even when he dug his nails into his fingers.[47]

Kawakami called this extraordinary experience the first great death (*taishi ichiban*), a Zen Buddhist term for the loss of the small self or ego.[48] Some Japanese writers, however, treat Kawakami's experience as the result of an epileptic seizure.[49] They point out that the physical symptoms he described may well have been manifestations of an epileptic attack. Still others may prefer to interpret this event as a reaction to extreme psychological distress, a case of nervous fatigue leading to an emotional breakdown. All these interpretations may have had some bearing on this unforgettable moment in Kawakami's life. Like Martin Luther's fit in the choir, which was also characterized by partial unconsciousness and a loss of muscle control, Kawakami's strange experience may have been on the "borderline between psychiatry and religion." Erik Erikson, in his book *Young Man Luther*, argues that Luther's "epileptoid paroxysm of ego-loss" could be conceivably interpreted in one of several ways: "as one of a series of seemingly senseless pathological explosions; as a meaningful symptom in a psychiatric case history; or as one of a series of religiously relevant experiences."[50]

Kawakami himself clearly felt that his strange experience had the character of a religious revelation. In the course of his solitary meditation, he had penetrated to the origins of his unhappiness and anxiety. He had discovered that his desire for success and fame, and his consequent fear of failure, was the cause of his suffering, and he had recognized the paradoxical truth that, to find himself, he had first to lose himself.

Illness, unhappiness ... are proof that you are selfish and egocentric; in order to root out your selfish nature, Heaven visits such afflictions upon you. We must really call them blessings. These so-called afflictions, in short, are means to lead you into the realm of enlightenment. In the end, it will come about that you shall be given absolute peace and well-being.[51]

Such religious experiences were part of Kawakami's cultural heritage and not at all uncommon in the history of East Asian religion. Kawakami's mystical encounter bears a remarkably close resemblance to that of Wang Yang-ming (1492–1529), founder of the Chinese Neo-Confucian School of Mind, whose great enlightenment had also come when he was alone, at night, exhausted after days of meditation on the question of how to become a sage. In Wang's case too the experience had involved giving up "all thought of personal success or failure, honor or disgrace" and had occurred at a time of great personal crisis.[52]

The keystone of these two cases of mystical experience was their ethical orientation. In both cases, the religious seeker was directed toward "action on behalf of all mankind." One sees this Confucian impulse toward socially significant ethical behavior rather than ascetic withdrawal in the meaning Kawakami drew from his "first great death." At the moment of enlightenment, his perplexing *kōan* was solved. All along he had been considering "this five feet of body" as his own possession, but his body was not his own, either to protect or to mortify. Rather, it had been entrusted to him as a public vessel (*tenka no kōki*). He must care for his physical being, because it was to be used thereafter in the service of mankind. It was not necessary, therefore, to renounce all desires, but only selfish desires. The problem was how to cultivate his private self so that he might become "truly worthy of being a public servant."[53] This was an expression of the time-honored Confucian notion that self-cultivation must precede social reform and that only the man who has first put his own home in order can "rule the realm."

After the experience of the "great dying" and the emotional letdown following the crisis, Kawakami's evangelical drive faltered. "Why, I don't know. Perhaps you can say because I came to have nothing at all. I lost the will to proselytize independently, yet ... the will to pursue economic research did not immediately awaken." It was "in the natural order of things" that he decided at last to become a member of the brotherhood of the Garden of Selflessness, a dramatic move for those who observed from the outside, but an anticlimax for him.[54]

The Garden of Selflessness phase in Kawakami's life was short-lived but instructive. He moved to the community fully prepared to stay.

Giving up his apartment and selling all his books (with the exception of Tōsons's *Young Greens*), he arranged to support himself by writing a new series of articles for the *Yomiuri* newspaper called "Trends of Life" ("Jinsei no kisū"), and at the beginning of January 1906 he built a house near the grounds of the main hall.[55]

Although Hajime lived with the community less than two months, his brief stay provided him with the moratorium he needed to clarify his own confused thoughts. While under Itō's tutelage, he tried to relate the teachings of selfless love to economic behavior. At first he succumbed to Itō's persuasive defense of all acts, including economic behavior, as natural and good, because they were part of the universe of selfless love. Viewed in such a light, he could justify his own career in economics because the natural laws and workings of the economy were in accord with the activity of selfless love. After all, it was natural for men to act as they did; they agonized over right and wrong only because they did not realize that their fate was entirely controlled by natural forces.[56]

Before long, however, Kawakami comprehended the philosophic gulf separating himself from Itō Shōshin's teachings. Kawakami's image of the model man combined the Christian spirituality of Tolstoy with the practical activism of Toynbee; Itō did not fit this description, because his teachings denied all distinctions between right and wrong, good and bad, and discouraged all efforts to change the world. Itō preached selfless love, demanding that his disciple accept the world the way it was. The young Kawakami, on the other hand, was seeking first to govern his passions and regulate his behavior and then to reform the social order. He was trying to bridge the gap between his ideals and the reality of Meiji Japan; Itō denied the gap existed. Selfless love was not the same as absolute unselfishness, and Itō Shōshin was no Christ.[57] Recognizing his mistake, Hajime withdrew from the sect on February 6, 1906, returned to Tokyo, and sent for his wife and child. Fortified by his religious experience, presumably purged of egoism, he was now prepared to resume his life in society.

4. the way in the modern world

When Kawakami Hajime left the Garden of Selflessness to resume his career, he discovered that he had become something of a literary hero among his intellectual contemporaries. Yoshino Sakuzō had commented favorably on his translation of Seligman's book,[1] and the socialist organ *Yorozu Chōhō* had referred to him as talented.[2] With the publication of his *Critique of Socialism* during the latter half of 1905, Kawakami's name had become familiar to a wide newspaper-reading public, not in the least because of the dramatic revelation of his identity. The many readers who had followed his provocative series of articles on socialism and had then read his emotional letter of resignation could hardly be expected to forget his name. The twenty-seven year old man who had renounced the goal of fame found, ironically, that he had already become famous.

Fame did not guarantee a steady income, however, and having discontinued his articles for the *Yomiuri* newspaper when he left the Garden of Selflessness, Kawakami was once again unemployed. In the early months of 1907, he decided to return to journalism, this time publishing his own periodical, the *New Magazine of Japanese Economics* (*Nihon keizai shinshi*). Kawakami apparently took over the magazine from his former university professor, Matsuzaki Kuranosuke, who had originally established it in cooperation with Prime Minister Katsura Tarō to air governmental problems arising at the end of the Russo-Japanese War.[3] In undertaking to continue the magazine, therefore, Kawakami publicly associated himself with the nationalist economic thought which Matsuzaki represented. This orientation was evident from his opening statement in the first issue, significantly published on the anniversary of the Meiji Emperor's accession to the throne:

> When I look at the treatment of economic problems in the publishing circles of various countries, I see that a variety of opinions appears to have mushroomed, without any single agreement. Some, in extreme cases, have advocated individualistic, laissez-faire economic activities as ideal, while others are one-sided and try to control our national economic policy in the interests of a certain class. These are not appropriate for our country. Therefore, on my own, I venture to publish the *New Magazine of Japanese Economics,* in order to solve all contemporary problems in the interests of the nation.[4]

In order to understand the appeal of nationalist economics to Kawakami, it is necessary first to locate his writings within the broader context of Japanese economic thought at that time. What other approaches to economics were available, why did he choose nationalist economics, and what did this particular school of economics mean to him?

Social Policy

Two schools of economic thought existed in the late Meiji period. The advocates of each school, economists both in and out of the academic world, differed largely in their assessment of what role the government should play in the regulation of the economy. Taguchi Ukichi's *Tokyo Economics* magazine (*Tokyo Keizai zasshi*) was the mouthpiece of the laissez-faire school, and the *National Academic Society* magazine (*Kokka gakkai zasshi*), published by members of the law faculty of Tokyo Imperial University, represented the national economic or social policy position. Until their newspaper closed down in 1907, the socialists had constituted a kind of third camp.[5]

Championing social policy was a small group of Tokyo law professors gathered about the person of Kanai En. Kanai had studied in Berlin under Roger Wagner, one of the leading figures of the German school of social policy. He returned to Japan in 1890, announcing his enthusiastic support of Wagner's theories. From that time on, members of the law faculty had devoted their lectures on economics to expositions of the German historical school's approach to political economy. These were the lectures Kawakami had heard when he attended the university.[6]

The concerns of the German political economists were eminently practical—contemporary problems, social reform, and administrative techniques. Their major objective was to train young students for bureaucratic posts in the German state. The German conception of political economy was a subject grounded in national history and directly related to the problems of governing. Consequently, German scholars favored the historical method, with its attention to the details of one nation's history, over the bare-boned, analytic approach offered by the founders of English economic thought. Germans felt that Englishmen, in their eagerness to discover natural and universal laws, too often pushed aside those particular events in one's own country's history which were the very stuff of national life.[7]

The German and British approach to economics differed in another way. However intent the classical economists may have been on establishing a science of economics, their "natural laws" led to certain political conclusions. If the slogan "government is best which governs least" became the watchword for a century of English liberals, it also became the touchstone for generations of German critics. Proponents of a strong state, the Germans were understandably opposed to the English doctrine of laissez-faire. In its place they supported an active role for the government in the economic affairs of the nation.

To characterize the German approach to economics, Joseph Schumpeter once told a joke about the German who asks, "Economics, what's that? Oh yes, I know ... you are an economist if you measure workmen's dwellings and say that they are too small."[8] Economic historians like Brentano and Wagner played dual roles as statesmen and scholars, feeling equally responsible to society and to scholarship. "Lujo Brentano addressed his classes as he would have political meetings, and they responded with cheers and countercheers."[9]

Members of the German school did not thereby consider themselves any less scientific, however, for they saw science as implying the scrupulous handling of research materials. In fact, research was synonymous with science. The voluminous studies in what was essentially economic history which came out under the direction of leading members of the historical school were models of accuracy and thoroughness. Moreover, the German historical school was not entirely uninterested in economic theory. Monograph topics dealing with specific problems in institutional history were farmed out to seminar students year after year, in the hope that after several decades, enough data would be collected to form the basis upon which generalizations could be made about economic phenomena.[10]

Sincere though their intentions were, German academics have often been criticized for being agents of the German state. The kinds of social reform measures they suggested and Bismarck enacted were calculated to deflate a growing German socialist movement. This fact explains part of the appeal which *Sozialepolitik* had for the Tokyo Imperial University professors. Their interest in warding off socialist movements made them sympathetic to the German school. But more important reasons for the appeal of German economics were, first, the prestige of German thought in general, and second, the similarities between Meiji statecraft and German social policy doctrine. Japanese economics professors shared with their German counterparts a similar appraisal of their own political roles as policy advisers to the nation. Disturbed by labor problems within the industrializing nation, they had pledged their full support to the study of social policy. A handful of men, led by Kanai and Kuwata Kumazō, established in 1896 a small study group which became known as the Association for the Study of Social Policy.[11]

The members of the association, like their German counterparts, stood equally opposed to laissez-faire doctrines of unbridled economic competition and to socialism. Denouncing both liberal economic and socialist solutions to the economic problems of the day, the association took a stand half way between the two: their prospectus proclaimed their hopes to preserve the private enterprise system, but to prevent, through the exercise of state power and the responsible activity of individuals, the class friction that seemed to be a concommitant of that system. Their ultimate goal was the reconciliation of labor and capital in the interests of social harmony.[12]

Over the years, membership in the association grew to include scholars from universities all over the country, lawyers, industrialists, and "progressive" government officials. For twenty-four years, beginning in 1900, the association held an annual conference at which members were invited to read their studies of important contemporary social and economic problems. Papers were presented on the need for railroad expansion or the conditions of the small farmer, or on workers' insurance and the need for factory laws. The association's annual report on the conference (*Shakai seisaku gakkai kaihō*), which contained these papers, became a repository for the most authoritative knowledge in the nation on economic affairs.[13]

This is not to suggest that social policy was synonymous with social welfare, or that the association's involvement with labor problems was strictly motivated by sympathy for the workers. It was not. If the social policy proponents concerned themselves with reform measures, it was only to ensure that social discord would be prevented. To build a strong and prosperous nation was the goal implicit in every scholarly paper on workers' insurance. Platitudes about national harmony and the need to reconcile the conflict between workers and capitalists concealed a growing fear of the union movement. Outbreaks of strikes after the Russo-Japanese War confirmed these fears.

While it would be wrong, therefore, to cast the members of the social policy camp in the role of democratic reformers, it would be equally inaccurate to overlook their efforts entirely. Until the rise of the democratic and socialist movement after World War I, the Association for the Study of Social Policy served as the only respectable forum in Japan for the discussion of contemporary social problems. Its conference rooms and publications provided a stage from which to discuss these serious conflicts in Japanese society that had arisen during the course of the nation's rapid modernization.[14]

In fact, in the early years of the association's existence as a small study group, members had run the risk of being mistaken for socialists. In their advocacy of factory laws and their eagerness to disseminate information on labor problems, they gave some people the impression they were part of the labor union movement. Actually, they were vehemently opposed to Katayama Sen's efforts to organize a labor program. When the Socialist Party was dissolved in 1901, members of the association, eager to keep their position separate in the public eye from the socialists, issued a statement emphasizing that social policy was not to be confused with socialism.[15]

In 1905, when Kawakami Hajime was young, polemical, and anonymous, he had railed against the "epigones of bureaucratic scholars" for their cowardice in rushing to disassociate themselves from the socialists. Doubting that a clear-cut distinction between socialism and social policy existed, he had argued that the two were different only in degree.[16] Once again in 1918, on the eve of his turn to Marxism, he would attack the aims and methods of social policy. In the years in between, however, he, too, relied on social policy to preserve national harmony and rectify the wrongs of the capitalist system. We can see this approach to economics in the magazine he put out beginning in 1907.

The Conflict between Ethics and Economics

The central focus of Kawakami's *New Magazine of Japanese Economics* was the capitalist economic system. Heralded as a panacea for backward Japanese society, industrialization had turned out to be a mixed blessing. The young Kawakami's naive enthusiasm for economic development, expressed as early as his eleventh year in his school essay urging a policy of wealth and strength, had been steadily eroded in his university years by mounting evidence of imperfections in Meiji society. Not surprisingly, Kawakami now confessed to having ambivalent feelings toward industrialization.

On the one hand, the young scholar of economics was exhilarated by the material progress which industrialization had brought to Japan. The whole mood of the Meiji period was one of almost breathless change; but far from fearing the relentless forces of technological progress, Kawakami observed the "trends of the times" with wonderment, welcoming such items of mechanical ingenuity as the pocket watch that inundated Japan. "Just think of it," he wrote, "things we common folk would hardly dream about, such as 'sky travel,' will soon be possible."[17]

Industrialization was thus more than a necessary evil. The invention of machines was the single most important reason for the great material progress of the contemporary world; the industrial revolution, based on machine technology and division of labor, had not only preserved the independence of Japan, it had also raised the nation to a higher level of material well-being.

Yet, two factors diminished Kawakami's delight in Japan's progress into the modern world. His discovery of the first problem, the "collision between economics and morality," had led him to abandon his teaching position. Kawakami had grappled with this conflict between economics and ethics on a personal level; he was now raising it to the level of a general national concern.

The second problem, the collision between agriculture on the one hand and commerce and industry on the other had also occupied him during his years as an instructor in agricultural economics and was a favorite subject of university professors and government leaders. These two problems, he realized, were intimately related—crises of a society in upheaval.[18] For the first time, Kawakami began to explore these crises in their historical context.

"Long ago we said samurai, farmer, artisan, and merchant, and the merchant was the most lowly fellow . . . In the present world, however, . . . the merchant spirit is in style . . . Whenever anyone opens his mouth, the first thing he asks is not whether something is right or wrong, but whether or not it is profitable."[19] With these words, Kawakami succinctly articulated the reasons for his misgivings. Capitalist economic organization was commercializing the entire economy, placing all social relations on a money-making basis. His father had spent most of his life in the confines of the four-class structure of Tokugawa society, where the merchant occupied the lowest rung of the social hierarchy. Not only was the maligned merchant liberated by the Meiji industrial revolution, but as Kawakami expressed it, everybody had become some kind of merchant.

The second unfortunate byproduct of industrialization was the impoverishment of the agricultural class. Kawakami complained that the government concentrated on the promotion of manufactures and trade, to the neglect of the farmer. With young people leaving their homes to make their way in the big cities, Kawakami predicted that the farm population would decline, and he worried that agriculture would disappear altogether.[20]

It was indeed true that farm productivity, especially after 1905, had failed to keep pace with the steadily increasing rate of production in the modern factories. The gap in productivity and in standard of living between the traditional and modern sectors of the economy had become conspicuous after the Russo-Japanese War, when modern industry, greatly stimulated by the war, had begun to develop more rapidly than the rural economy.[21]

But, if modern industry represented progress, why was Kawakami so concerned about the fate of Japanese agriculture? Did this eagerness to strengthen the agrarian economy reflect a wish to preserve at least a part of the Tokugawa way of life? Some leaders in Meiji society had actually expressed such a sentimental attachment to the undisturbed countryside, voicing their dismay over finding lemonade being sold in a provincial store or farmers purchasing their umbrellas instead of making them by hand.[22] A cursory glance at Kawakami's two books on agricultural economics might leave the reader with the impression that he, too, was reluctant to part with the past, but this was not exactly the case.

In the two books which he published immediately before and after entering the Garden of Selflessness, as well as in some of the articles in his *New Magazine of Japanese Economics,* Kawakami seemed to be breathing new life into the old "agriculture is the base" theories (*nōhonshugi*) on which all of Tokugawa economic thought had rested. This physiocratic bias appeared in the pages of his *Reverence for Agriculture in Japan (Nihon sonnō ron)*, where he spoke of agriculture as a basic industry and commerce as merely derivative.[23] It appeared in his *Agricultural Management in Japan (Nihon nōseigaku)*, where he argued for the protection of agriculture.[24] A romantic nostalgia for the self-sufficient life of the past, however, did not afflict Kawakami, because he viewed social change from a historical perspective and equated economic development with progress:

> As a child I grew up in a castle town in the backwoods of Yamaguchi Prefecture. At that time, as I recall, you could invariably find in almost every home implements for weaving and spinning. With these tools, people spun yarn from the cotton and with the yarn they made what we call hand-woven cloth. In this way they satisfied their everyday clothing needs. When you said weaver or cook, you meant the elementary accomplishments that no good daughter-in-law should be without.
>
> When I returned home recently, however, I found they had put all the equipment for spinning and weaving in the storage room. It is not that the people back home have lost their industriousness; it is simply that everywhere economic organization has completely changed, owing to the division of labor.[25]

In calling attention to the lag in the farm economy, Kawakami had not meant to deny the nation's industrial goals; quite the contrary, his very concern with these goals had prompted his defense of agriculture. A quotation to this effect from Edward Gibbon opened the introduction of the second part of *Agricultural Economics:* "Prosperous agriculture means for us prosperous manufactures, and from an economic point of view the interest of the plough and the loom are identical."[26]

Industry could not advance at the expense of agriculture, as government leaders were inclined to believe; agriculture, industry, and commerce had to develop at an equal pace. What Kawakami actually was advocating was not the preservation intact of the traditional agricultural economy, but its modernization: he urged the government to invest in the mechanization of agriculture. Against those who contended that agriculture was a bad risk, subject to the whims of nature, Kawakami pointed out that farming was losing its original characteristics. Machinery could free the farmer from his dependence on natural forces. To those who said agriculture was unprofitable, he retorted that they knew nothing about the experiences of other countries. A revolution in agriculture is taking place, he told them, citing the examples of Belgium and the United States.[27]

He further argued that there were concrete economic advantages to be gained by developing agriculture. Money in the hands of the farmers would increase the demand for manufactured goods. The increased demand would in turn stimulate industrial production, while stopping the flow of farmers to the cities. Factory wages would rise as a direct consequence. A decline in farming, on the other hand, would force more of the rural population into the cities, causing wages to drop, as the cities swelled with job seekers.[28]

Kawakami saw other dangers involved in making agriculture the stepchild of the government's industrialization efforts. The farm economy served military as well as economic needs. It was conceivable that in times of war, food imports might be jeopardized. Since economic independence was the basis of national independence, the promotion of domestic food production contributed to the nation's military preparedness.[29]

In further support of the military argument, he offered the opinion that peasants make the best soldiers. It was a curious argument for one whose family, only thirty years earlier, had enjoyed elite warrior status, but he reasoned that for one thing peasants were healthier than others and for another they were more docile and obedient. "Agriculture is the well-spring of a strong army. The preservation of agriculture is necessary for victory in war."[30]

If the paddy fields supplied the front lines, they also supplied the assembly lines. Since the urban death rate exceeded the birth rate, he argued, one would expect the city population to decline. This was not the case, however, because of the constant flow of new men from the countryside where the birth rate was fortunately higher than in the cities and unlike the cities, did not exceed the mortality rate.[31]

There were also humanitarian considerations motivating Kawakami's opposition to total industrialization. He wanted to promote small-scale rural industry so that families could stay together. Parents could then supervise their children and women would not be mistreated by employers.[32]

It is evident that Kawakami tried to marshal every conceivable argument to support a reverence for agriculture. These arguments were squeezed somewhat awkwardly into the Meiji goal of a rich and strong nation. Buried underneath this enthusiasm for the technological revolution, however, lay the abiding fear that if agriculture declined, the nation's solidarity would also diminish. Rural life preserved those values of personal obedience, frugality, and self-sacrifice prized in Tokugawa society and redirected in the Meiji period to the support of the nation. At stake in the clash between industry and agriculture was more than the weakening of an economic class. The new economic order threatened the very foundation of traditional social values, as the merchant spirit, liberated by Meiji Japan's industrial program, burst its confines and spread to all parts of the body politic. Masters of the profit motive, men committed to the new capitalist economic order, flouted their newly won success in the face of the virtuous farmer.

In Kawakami's idealized portrait, farmers were frugal, patriotic, obedient, and healthy. They made the best soldiers and the most loyal citizenry. Their closeness to the soil and their self-sufficient way of life had insulated them against the unhealthy tendencies of city dwelling. Above all, farmers did not work merely to increase their own profits, nor did they join labor unions to fight for higher wages. They were not extravagant, like townsmen, and they did not espouse revolutionary political ideas that threatened social tranquility.[33] The conflict between industry and agriculture was tied to a greater contest between the private profit system and Kawakami's altruistic ethics. "Reconciling the interests of the farmer, the merchant, and the industrialist is a large problem and, moreover, an exceedingly difficult problem ... It would be the greatest catastrophe to ignore morality for the sake of economics, and agriculture for the sake of commerce and industry."[34]

Yet, how could Kawakami hope to rescue old values through the modernization of agriculture, when capitalist techniques of mechanized farming would destroy the social basis of those values? His tortured logic captures the dilemma he perceived. He praised the evolution of an interdependent economic system based on the division of labor and he extolled the farmers' self-sufficient way of life. In his plans for modernization, he banked on the increased buying power of the agricultural class; and in his praise of agrarian traditionalism, he applauded the farmers' frugality. Farmers made good soldiers, but he hoped to make them good capitalists as well. He argued that the development of agriculture was economically profitable but then contradicted himself by calling the rescue of agriculture morally desirable precisely because the farm class traditionally scorned the profit motive.[35]

The dilemma Kawakami faced in formulating an economic policy that would "protect the intimate relation" between industrial and agrarian interests illustrates the difficulties inherent in trying to design a development plan capable of satisfying all sectors of an industrializing society. Robert Bellah remarks, "It is possible that the continuous rationalization of the economy may create such strains in the social system as a whole that the values legitimizing a rational economy may be threatened."[36] Industrialization, seemingly bound to the profit motive, stood on one side, and agriculture, symbolizing moral values, stood on the other side of the great divide that split the Japanese nation in the modern world; and Kawakami was himself torn between the economic rationalism that would make the nation great and the economic individualism that might, he feared, destroy it. Gaps between rich and poor, between industry and agriculture, between economics and ethics, and ultimately, between the West and Japan—a never-ending nightmare of antagonistic forces—threatened national unity at every turn. How could these conflicts be resolved?

The ideal of social harmony, expressed in a word which Kawakami frequently used, harmonization (*chōwa*), may have recalled the near-distant past, or that idealization of it which the tumultuous new era had provoked. But backward glances at Tokugawa agrarian society did not divert Kawakami from his forward-looking policies. He could not restore a self-sufficient economy, nor did he want to; capitalism was the instrument of material advancement. He hoped instead to alter the ethical component of the capitalist system, by replacing the ethics of self-interest with an ethics of altruism. The question he raised was whether the capitalist system could work without economic individualism. Was it possible to apply his newly formulated moral standard to the remedy of Meiji Japan's problems?

The Reestablishment of Harmony: Confucian Nationalism

The pursuit of self-interest formed the basis for the classical economists' rationale of the private enterprise system. In his formulation of the principles of capitalism, Adam Smith argued that the private profit motive stimulated men's drive to invent better products and improve methods of production. The belief that individuals worked best when working for their own interests was basic to the justification of capitalism. "It is not from the benevolence of the butcher, the brewer, or the baker that we expect our dinner," wrote Adam Smith, "but from their regard to their self-interest. We address ourselves, not to their humanity, but to their self-love, and never talk to them of our necessities, but of their advantages."[37]

What then prevented individuals from trampling over each other in their grim struggle for profits? What produced the rational order and harmony which, according to the classical economists, was inherent in capitalism? The answer to these questions rests in the mechanism of competition itself. If the tradesman cheats his customers, he will soon lose out to more honest businessmen; if he pays his workers less than other employers do, they will go somewhere else. The market was a self-regulating system operating according to its own laws to provide for the quality and also the quantity of goods and services required by society's members.

Kawakami remained unconvinced that men were morally justified in snatching at opportunities to enrich their material existence, even if such behavior yielded social benefits. There was an alarming number of dissonant notes in the presumably harmonious capitalist society to confirm Kawakami's suspicions that the self-serving principles of classical economics provided neither the proper ethical basis for society nor a minimum livelihood for all. The gloomy predictions of Malthus still haunted him as he searched for ways to ameliorate the ills of capitalist society without impeding technological progress.

One might argue that Kawakami was beating a dead horse. Few people in Meiji government or business circles actually championed the philosophy of laissez-faire economics. For one thing, the Meiji government actively intervened to promote industry and, for another, both government officials and business leaders rationalized business activities in terms of national interests, rather than private gain. Even the most successful entrepreneurs, like Shibusawa Eiichi, claimed they were working for the good of Japan and not for their own enrichment. Industrialization in Japan from the very beginning was fortified by an ideology that drew on pre-industrial Japanese values; although a prominent business class emerged in the Meiji period, its members, latter-day samurai in business suits, preferred to dress their commercial endeavors in the cloak of patriotic nationalism.

Nevertheless, there were signs toward the end of the Meiji era that some individuals were indeed profiting from industrialization and that crass materialism was becoming at least for some a way of life. While imperial edicts warned against the evils of the time and morals courses in the nation's schools drilled Japanese children in the virtues of thrift and selflessness, an increasing number of people appeared more concerned with their own material prosperity than with the nation's well-being. The spending sprees of the nouveaux riches and the gambling craze among wealthy women furnished proof of the moral laxity of the age.[38] Such gaudy displays of wealth and openly hedonistic spending seemed to spell the triumph of commercial values.

At the same time, other members of society, like farmers and industrial workers, had fallen victims to poverty. The clamor for labor unions and factory legislation begun over ten years earlier and supported by the small but vocal socialists around the turn of the century, had raised the spectre of violent revolutionary means to redress injustices. In the light of these ideological rumblings it was difficult to sustain a theory of natural harmony.

Against the background of spreading commercial values and economic unrest, Kawakami's efforts to reconcile money-making and morality assumed increasing urgency. The fear that private material gain would replace moral pursuits as the goal of human life dominated his thinking and was also reflected in his ascetic life style: forsaking the indulgences of his university school days, he abstained from alcohol, lived a simple life with few diversions, and if his autobiography does not lie, he allowed himself only two gustatory pleasures, *manju*, a Japanese sweet cake, and pipe tobacco. Kawakami's aristocratic disdain for money matters—and perhaps, too, his fiscal irresponsibility—led Hide to insist on managing the family's domestic finances, a function most Japanese wives today perform, though possibly with less rigor. The knowledge that his wife took his entire earnings, leaving him only tobacco money, lends piquancy to the contradiction between the economics professor's careful avoidance of cash transactions and his earnest efforts to comprehend the operation of capitalism.

Yet, in this very paradox one finds clues to Kawakami's early thinking about economics. Capitalist economic behavior—man as buyer and seller—was not something he took for granted but, rather, something he first had to justify on ethical grounds. His initial economic studies, therefore, scarcely concerned economics at all, because he was not so much interested in how capitalism worked—the laws of supply and demand—as in how capitalism could be legitimized.

This inclination in his thought helps explain why he turned for guidance to Tokugawa neo-Confucian scholars for whom the conflict between money making and moral cultivation had also posed a basic dilemma. Kawakami's awakened interest in Tokugawa economic thought around 1907 coincided with the beginning of a general revival of interest in the Tokugawa political economy among Meiji Japanese scholars and typified efforts on the part of some Japanese intellectuals and government leaders, in Kawakami's words, "to explain the relation between economics and morality" and to "harmonize Eastern and Western culture."[39] These attempts to find in the native Japanese intellectual heritage a persuasive philosophical alternative to the Western ethos of enlightened self-interest stemmed from the tendency to see all of Western culture as solely materialistic. Conspicuously absent in Kawakami's writings is any mention of Christianity as a source of public virtue in the West. Even though he had himself been influenced by

Christian ethical tenets, Kawakami overlooked the possible value of Christianity as the altruistic side of Western morality. He also failed to distinguish between the self-restraint and "sober virture" required of the successful entrepreneur and hedonism. This inability to find any redeeming virtues in the West, either within the ethos of capitalism or outside it, in the realm of religion, prompted his research on Japanese Confucian thinkers.

Confucian scholars' efforts to contend with the growing money economy of Tokugawa times made their writings surprisingly relevant to Meiji society's ailments. Kawakami found the works of the eighteenth century scholar Kumazawa Banzan especially pertinent in this regard. Banzan, too, had wondered how to reconcile morality with the deliberate pursuit of private profit. His solution was to separate self-interest from economic interest. "Banzan does not absolutely renounce all . . . desires," Kawakami concluded, "he renounces only those which are the wrong desires."[40]

The problem of human life, according to Confucian teachings, was how to transcend "wrong desires" for personal gain beyond what was necessary to live, and have only "right desires," to benefit society. "Men must desire and must not desire," wrote Kawakami, quoting Confucius. Human life was a journey toward the goal of moral perfection, the Way, when one wanted to do what one ought to do.[41] This life quest presented the greatest challenge to men in the present age, when they were dependent upon money-making enterprises and exposed to commercial values. Was it possible to "pursue righteousness and throw away profit" and still survive in the modern world?

This question formed the basis of Kawakami's life and thought. His solution in the late Meiji period was to link Confucian moral cultivation to patriotic loyalty: the individual transcended the selfish aspects of the capitalist system even while operating within its framework by denying his own needs and working for the greater good of the nation. Hard work and frugality became a patriotic Japanese equivalent of the Calvinist concept of stewardship, with economic activity morally justified only when performed with the interests of the nation in mind.

In this solution Kawakami redirected Confucian altruistic ethics to the service of the nation, using the national good as the rationalization for economic behavior. He found such a solution acceptable, however, only because he believed, with the Confucians, that government was (or should be) a moral agent mediating the welfare of all the people. This assumption was implicit in his resolution of the conflict between economic acts and selflessness:

If the producer has the firm belief that he is making a profit for the sake of everyone, then according to that standard, he will manage his business enterprise, and even if economic organization is left as it is today, all enterprise, while under the name of private enterprise, will actually be government enterprise, and persons called producers will actually be public servants.[42]

He thus purified economic activity by translating it into a branch of government service. The moral standard which Kawakami had established for himself he now advanced for the rest of society: all men should act as "public vessels."

What did this social morality mean in terms of everyday behavior in the economic world? Like Shibusawa, the successful Meiji entrepreneur who boasted of combining the *Analects* and the abacus, Kawakami seemed to imply that profit making was legitimate as long as the businessman pursued profits for selfless reasons. He may have meant that the wealthy should avoid a life of extravagance and workers refrain from agitating for higher wages, but he did not elaborate on these details, nor did he find it necessary to demonstrate how frugality, ethical integrity, and "firm beliefs" would solve economic problems.

Kawakami's silence on these critical aspects of economic theory was significant. His neglect of economic analysis and his reliance on moral reform instead to cure economic problems reflected the traditionalist bent of his thought. Frugality was the favorite remedy of Tokugawa economic thinkers charged with balancing supply and demand. "Let there be activity in the production and economy in the expenditure," advised the *Great Learning*. "Then the wealth will be always sufficient."[43]

The Confucian influence on Kawakami's thinking was also seen in his view of government as a moral agent regulating the economic behavior of individuals. A fixed plan was necessary to harmonize competing economic interests, he wrote; the private enterprise system was inconvenient and not conducive to national unity.[44] Kawakami preferred to entrust the task of reconciling economic conflict to moral statesmen whose policies rose above the contending play of individual desires. Social policy was ideal in this sense, because it offered an approach to economics that closely approximated Kawakami's own intellectual proclivities, blending as it did Confucian ideas of statecraft with modern economic research.

A Member of the Establishment

These Confucian-tinged nationalist views were strongly in Kawakami's favor when his name was mentioned in 1908 as a candidate for appointment to the faculty of Kyoto Imperial University. It was a prominent position. Although the Kyoto campus was only eleven years old, as the nation's second Imperial University it was outranked only by Kawakami's alma mater in Tokyo. This reputation was probably based as much on its order of establishment and governmental affiliation as on its academic quality: treated with the same high regard reserved for top-ranking public officials, faculty members in the handful of publicly financed Imperial Universities before World War II and especially those in the oldest ones enjoyed an impressive degree of status.

Although the invitation to join the university came as a complete surprise (he was asked to replace a faculty member who had suddenly become ill), Kawakami accepted the offer with unaccustomed speed. It was one of the few decisions in his life that he made without hesitation. He welcomed the appointment not only for the prestige it conferred but also because it freed him from having to publish his economics magazine. Kawakami had come to realize that he had overreached himself in attempting to defend nationalist economics against the laissez-faire arguments of such well-known writers as Taguchi Ukichi, whose *Tokyo Economics* magazine had served as a foil for Kawakami's enterprise. Lacking confidence in his own scholarly background as an economist, Kawakami felt grateful for being "picked up by Kyoto Imperial University."[45] He hoped the faculty appointment would provide him with the time he needed to explore the science of economics, a subject in which he was admittedly still a novice. Excited by this unexpected good fortune, he scraped together the small amount of money at his disposal for moving expenses and in 1908 brought his family to the more leisurely atmosphere of the old imperial capital to begin a long and productive academic career.

Separated from Tokyo, the political heart of the country, Kawakami spent the next two decades of his life on campus and in his study, his reformist ardor seemingly spent. He rapidly climbed toward the top of the academic hierarchy, receiving the coveted Ministry of Education scholarship for study abroad in 1913 and, upon his return two years later, the rank of full professor and the title of doctor. If he had any qualms about retreating from social reform to the academic world, he never expressed them. The years after his appointment to the Kyoto faculty were the most tranquil of his entire life and probably for this reason, he devoted few pages to them in his autobiography. He must surely have noted, however, that he joined the faculty on his thirtieth birthday when, according to Confucius' stages of spiritual development, a man "planted [his] feet firm upon the grouund."[46]

Comfortable in his professorial role, Kawakami devoted all his energies to scholarship, adopting a detached writing style that demonstrated his growing familiarity with the developing social sciences and the methods of modern scholarship that were entering Japan at this time. Behind his devotion to academic pursuits, however, lay his abiding interest in the problem of selfishness. After joining the university, he became interested in studying economic theory and especially the operation of the capitalist system—the laws of production, the origin of profit, the creation of wealth. Above all, he wondered how various European economists treated the problem of economic individualism: did they consider the unlimited satisfaction of private interests to be an unavoidable concommitant of industrial production? Although he had left behind his youthful involvement in religious movements to resume his economic studies, these very studies were influenced by his ethical concerns: "The question of the right and wrong of individual, selfish behavior became without my knowing it the basic theme of my thought."[47]

Over the years, Kawakami forged a reputation as an authority on bourgeois economics, but his research and lectures actually ranged over a broad area of learning, not at all or even mainly confined to economic studies. He published articles on Cro-Magnon man, primitive methods of trapping elephants, and ant society. In his first years in Kyoto, he also translated several books on economic theory, including the American economist Irving Fisher's *Capital and the Rate of Interest* and N. G. Pierson's *Theory of Value,* and he followed the work of the Austrian economist Böhm-Bawerk. All of them were engaged in establishing a purely analytic economics.[48]

Until he became a full professor, however, Kawakami was required to teach economic history and the history of economic thought, as well as several policy-related courses on traffic and agricultural management, while his senior colleague, Tajima Kinji, lectured on economic theory. It was not until 1916, when Tajima began taking turns with him teaching economic theory, that Kawakami had the opportunity to coordinate his research interests with his lecture preparations and by that time his own ideas were already under attack by other Japanese scholars, who were acquiring a Marxist framework of economic analysis. He did not touch on Marxism in his lectures nor for that matter did he publish any major work on British classical economics until after 1919, the year in which he began in earnest to study Marxism.[49]

After joining the Kyoto Imperial University economics faculty, Kawakami's name naturally became associated with the social policy school of economics. His own mentor, Matsuzaki Kuranosuke, was a disciple of Kanai En, the founder of social policy in Japan; and his colleagues, Tajima Kinji, Ogawa Gōtarō, Kanbe Masao, Kawada Jirō, and Toda Umiichi were all disciples of Kanai and Matsuzaki, as were members of the government bureaucracy, many of whom had studied under Kanai. Until his conversion to Marxism a decade later, Kawakami was closely identified with this illustrious group of scholars and government officials who called upon social policy to resolve social tensions in the interest of the nation.

Kawakami Hajime's advocacy of social policy might seem to call into question the sincerity of his humanist sympathies. His constant appeals to the overriding interests of the Japanese nation may even give the impression that he was a statist at heart. The following passage from his book *Economics and Human Existence (Keizai to jinsei),* published in 1911, is a typical example of his social policy position:

> The major difference between the West and Japan lies in the difference between individualism and nationalism. In the West, even socialism is based ultimately on the individual. This is also true of social policy in the West, which has as its goal the fulfillment of the individual. In our country, we can say that social policy has still not been realized. For example, even though we have moved toward establishing factory laws, the goal of these laws is not the reverence of the workers' personality or the increase of the class profit of the working class. As much as possible we consider [the workers] to be the tools of national industry. The improvement of these tools is necessary for the healthy advancement of the nation's industry. Thus, industrial social policy is not the social policy of recent times: briefly, if Western economic treatises concern distribution, Japanese economic treatises concern production. Looked at from the point of view of the individual, economics is oriented toward distribution; looked at from the vantage point of the nation, it is oriented toward production.[50]

How did Kawakami reconcile these sentiments on behalf of the nation with his sympathy for the welfare of the people? It is difficult to answer this question because Kawakami did not make a conceptual distinction between nation-state and society. As late as 1916 he wrote, "The nation is merely one form of society."[51] He thus tied together individual ethics, popular welfare, and state power in a unity impossible to disentangle. For this reason, it is often hard to determine whether Kawakami, during the years he espoused social policy, was a champion of the state or of the poor. He was actually both. His ethics of selfless devotion took as its object some reference group transcending himself; at this time in his life he did not seem aware of the possibility of conflict between the state and the people. Insofar as he expected the interests of the nation to be mediated by the state, however, his ethical stance became inseparable from ardent nationalism in support of governmental authority. A strong state was the antidote to selfish individualistic economic activity.

Such attitudes toward government and ethics may explain Kawakami's reluctance to embrace socialist solutions even though he shared the early Japanese socialists' distaste for the ethos of capitalism. He had been interested in socialism ever since his university days, when two prominent socialist critics, Kinoshita Naoe and Katayama Sen, had captured his admiration. He saw in the utopian socialist blueprint his own dream of a harmonious society of unselfish human beings and an efficient economic system. In his *Critique of Socialism,* in 1905, Kawakami had called the ideal socialist society a happy, beautiful, rational one, questioning only whether such utopian ideals could ever be achieved on earth. In 1909 he had approvingly described the Tokugawa thinker Satō Nobuhiro (1767–1848) as a socialist because Satō had recommended that the state manage all buying and selling and provide public relief for the poor.[52]

Despite his interest in socialism and his consistent defense of the socialists' right to free speech and free press,[53] there are several reasons why Kawakami resisted the appeal of socialism in this period before World War I. The first is the obvious reason that the Ministry of Education prohibited the teaching and publication of Marxist or even socialist ideas. The close connection between the faculty and government leaders would have prevented Kawakami from freely expressing his opinions, and his senior colleague in the hierarchy of economics professors on the faculty, Tajima Kinji, argued frequently and forcefully against Marxism. We know that Kawakami hesitated to defend socialist doctrines for fear of government reprisal. In 1905 a Tokyo Imperial University Law Faculty professor, Kosui Hirōto, had been forced to resign for publishing an article criticizing the terms of the Japanese peace treaty with Russia. Although Kosui's hard-line attitude toward the war represented the very opposite of the socialists' pacifist stand, nevertheless the issues of academic freedom and freedom of the press raised by the Kosui incident affected academics and intellectuals of every political stripe. Kawakami remained an ardent advocate of academic freedom throughout his career, but on several occasions, most notably in 1905, with the publication of his *Critique of Socialism,* he used an assumed name rather than run the risk of jeopardizing his career. Judging from several allusions to this problem of government interference, we can conclude that Kawakami deliberately refrained from publishing his honest opinions on such sensitive subjects as socialism or on the orthodox theory of the state (*kokutai*), a doctrine that, among other things, affirmed capitalism as the basis of Japanese economic organization.[54]

While constraints on academic freedom proved burdensome, however, these were not the major reason for Kawakami's rejection of socialism.

It becomes clear from reading Kawakami's autobiography, where he did feel free to express himself frankly, and from his writings in the period from 1908 to the first World War, that he did not understand fully the doctrines of Marxism and socialism or the differences between them, and that his approach to social problems, influenced by traditional Japanese views on reform, clashed with the modern tenets of socialist thought.

Kawakami was convinced that institutional reform should be preceded by individual moral reform. To perfect society, he repeatedly stated, it was necessary to reform both its material and moral sides. The material side corresponded to scientific or economic research and the moral side to philosophy or ethics. The socialists appeared to neglect this second side.

Another reason for Kawakami's aversion to the socialists in the late Meiji period was their apparent materialism. In clamoring for material gains alone the socialists seemed to be as selfish as the capitalists they sought to replace. Both economic doctrines were based ultimately on unbridled individualism. In addition, he felt that the socialists' encouragement of labor class activity created social disorders and class antagonisms, the very opposite of the social harmony he wished to preserve.

This criticism of the socialists highlights the difference between Kawakami's brand of humanism and theirs. He did not consider material well-being to be an end in itself but only the means for men to attain the real goal of human life—moral perfection. Secure from the struggle for sheer physical survival, men could concentrate on the cultivation of their spiritual life. On several occasions, Kawakami implied that the socialists, including Marx, were actually saying the same thing as Mencius. He thought he saw an affinity between economic determinism and the views of Mencius on the importance of economic well-being among the masses as a prerequisite for their education in morality.[55] To this extent, he found socialist thought compatible with his own ideas. In other words, Kawakami was drawn to those aspects of socialism which mirrored the Confucian version of the good society, but he was repelled from socialism at those points where its modern individualism conflicted with his own more traditionalist ethics of self-denial.

Repudiating both socialism and laissez-faire capitalism, Kawakami preferred to rely instead on government policies which, he believed, were formulated in the interests of the national collectivity. In the years before the upheaval wrought by World War I, he remained confident that such policies were possible, assuming an identity of interests between social welfare and nationalist goals. "Progress," he wrote in

1912, "has been accompanied by a gap between capitalists and workers. Class warfare is the struggle between capitalists and workers originating in this gap. Socialism instigates the struggle and social policy reconciles it."[56]

5. japan and the west

It was not surprising that Kawakami experienced a renewed appreciation of his own cultural heritage after moving to Kyoto; indeed, he may have discovered that rich heritage for the first time, for nothing in Iwakuni or Tokyo could equal the splendor of the old imperial capital. With its numerous landscape gardens and Zen temples, Kyoto preserved the essence of Japanese religious and esthetic values against the relentless push of change. Kawakami's daily stroll through the wooded Yoshida shrine grounds on the hill behind the university afforded vital closeness to nature, a momentary escape from the modern Western world which it was his professional task to comprehend.

Mellowed by the warmth of success—for clearly by the last years of the Meiji period he was securely established in the academic world— Kawakami settled down to a serious work routine. The rhythm of his professional and family life tempered the emotionalism of his student days, and his writings in this period immediately preceding World War I bear evidence of his new sense of confidence in himself and in Japan.

His professional life was going well. He continued to publish books and translate at a prodigious rate: in 1911 he published *Trend of the Times (Jisei no hen)* and *Economics and Human Life (Keizai to jinsei)* and the following year, he came out with *Studies in Economics (Keizaigaku kenkyū)*. In the same year, which marked the close of the Meiji period, he made his debut before the Association for the Study of Social Policy, presenting a paper on "Communal Life and Parasitism," and the following year he left for Europe and further study.[1]

His private life was also going well, though he and Hide were troubled by their young son's serious heart disease. Masao, who inherited his father's weak physical constitution, never regained his health; he died in 1926 after years of illness. With three children in the family (their other two were girls), Hide carefully budgeted her husband's salary and was able to put his book royalties into savings. Although strapped for money in their first few years in Kyoto, they always had enough for the daily necessities and lived comfortably in a two-story house near the university campus.

These were good days for the Japanese nation, too, basking in the victories of wars against China and Russia, acclaimed as a major power in East Asia, and proud of its remarkable industrial achievements. There were serious inequalities of wealth, it was true. There were also grave financial problems confronting government leaders: the nation lacked foreign currency reserves and faced a severe balance of payments crisis, solved only by the fortuitous outbreak of World War I. Moreover, a dangerous political contest had left the government stalemated in the 1911–12 cabinet crisis as the Satsuma–Chōshū clique fought to preserve its grip over Japanese politics. For the Japanese intellectual elite, who were sensitive to contemporary issues and felt personally responsible for Japan's fortunes, such trends were disturbing.

Yet for Kawakami, a leading member of that elite, these problems were only peripheral to his consciousness. Even the death of the Meiji Emperor in 1912, a sorrowful event that brought the curtain down on a glorious era for Japan, did not personally affect him. His faith in the nation and its form of government was probably stronger in these years than ever before or ever again. "Japan, ... has become great," he wrote. "Japan is a good country. The Japanese race is a good race."[2] This heightened nationalist sentiment sustained him until the end of the First World War, when his conversion to Marxism signaled the collapse of his optimism, just as Japan was beginning a turbulent new era in her history.

In these confident years between 1911 and 1915, Kawakami temporarily abandoned his economic studies, concentrating almost all his intellectual efforts on cultural comparisons instead. While barely concealing his preference for Japanese culture, his attempt to identify and at the same time affirm Japan's unique cultural traits was disciplined by the new objectivity that guided his studies. Kawakami's exposure to the new social sciences of anthropology and sociology instilled in him rigorous standards of analysis, which prevented his comparative study of culture from ending in rabid chauvinism. In this regard, he was not unlike one of his contemporaries, Watsuji Tetsurō (1889–1960), who was also "seeking a rationale for handling problems of comparative culture and ... for defending the necessarily Japanese nature of Japanese culture."[3]

What explains the fascination with cultural identity that dominated Kawakami's writings at this time? Although Kawakami's preoccupation with cultural differences suggests a new intellectual direction, it is possible that the German historical school's emphasis on adapting economic policy to nationality differences led him to explore more fully the nature of Japan's cultural distinctiveness. Moreover, his own negation of economic individualism in favor of a social policy geared to nationalist goals made his ethical formula for Japanese capitalist society different from the ethos of the West or, at least, different from England, France, and the United States; and the many ramifications of this basic cleavage between Japanese nationalism and Western individualism may have led him naturally to explore other nationality differences as well.

"What we reject," he wrote of capitalism, "is not the form but the spirit. What we reject is not the private ownership of industry, but the theory of private ownership, based as it is on individualism."[4] The ideas of rights and self-government, he explained, were products of that individualism; in Japan, where the concept of duty, rather than rights, prevailed, "we can only have bureaucratic government under nationalism." Nationalism, bureaucratic (as opposed to popular) government, and the sense of duty: these three aspects of national life distinguished Japan from the Western world. "Those persons who are respected in Japan are all dedicated to the nation. Herein lies the saying, 'We should revere the bureaucrat and despise the citizen.' "[5]

The heart of Japanese civilization, Kawakami further explained, lay in its special nationalism. In Japan the nation was an end and the individual the means, whereas in the West the individual was the end, and the nation the means. If the greatest morality for the Japanese was patriotism, for the Westerner, it was the development of the individual's personality.[6] Nationalism was not only the highest expression of morality in Japan; it was the religion of the Japanese.

> The god of Japan is the nation. Thus the emperor is the one who represents this sacred national polity [*kokutai*]. In other words, he concretely embodies the abstract national divinity. According to Japanese beliefs, therefore, the emperor is divine. That is, the emperor is a god [*kami*]. In the imperial constitution it is clearly explained that the emperor is sacred and inviolable . . . Thus, patriotism is the highest virtue in our country and, at the same time, patriotism is synonymous with loyalty to the emperor.[7]

This fundamental belief in the sacredness of the nation and of the emperor explained why Japanese and Western political theories concerning the nature of the ruler were diametrically opposed. In Japan, monarch and state were considered one and the same thing; the emperor was the divine representative of the state. In the West, on the other hand, the ruler was viewed as an organ of the state; he preserved his identity apart from the state. In a certain sense, everyone from commoner to king stood on an equal footing in Western society, because everyone had his own individuality, his own private life separate from his public duties. In government as well as in economics, dissimilarities between Japan and the West were rooted in opposing attitudes toward the individual.[8]

Kawakami's antipathy to the spirit of capitalism was thus followed by a general rejection of liberalism, based as it was on the affirmation of the rights and interests of individuals. Certain cultural differences, he seemed to be saying, could not be safely dissolved by the agent of modernization, because precisely these "special Japanese characteristics," when integrated into nationalist ideology, accounted for Japan's strength.

> Nationalism is the essence of our country. It is this nationalism that makes our poor, small country a capable, strong country. Look at ant society. With the exception of human beings, no other society has realized such a high degree of civilization as the ants. However, this is due to their society's nationalism. When we compare individuals [in the West and Japan], we Japanese are no match for the Westerners, either in wealth, power, knowledge, or physical prowess, but having once decided to cooperate and form the Japanese nation, we produced an able, great, and strong country because of this nationalist spirit.
>
> Therefore, I passionately desire the healthy development of this nationalism and I oppose all theories, movements, policies, and systems which are not useful or are even harmful to it, because it is the characteristic of the Japanese nation.[9]

It may very well be that Kawakami's preoccupation with comparative sociology was born out of a sense of discomfort with the Western model of development. Those aspects of Japanese culture which he identified as different represented areas of Japanese life stubbornly resistant to Westernization—as well they should be, in his view—and hence in conflict with Western ideals. The question was whether Japan, in the face of persistent Western influences, and especially under the impact of scientific thought, could continue to maintain its special nationalism, based on its cultural distinctiveness. This was a question which few Japanese felt they could ignore.

In his role as an interpreter of Western thought, Kawakami experienced most poignantly the conflicts inherent in the introduction of modern scientific rationalism into Japan. Precisely because the Japanese defined their uniqueness in terms of the mind or spirit of the people—a corpus of ancient myths, beliefs, and values—their national unity was threatened by the invasion of scientific consciousness. "Popular beliefs," he explained, "are no more than old knowledge that has accumulated and jelled, becoming emotion." Such old knowledge collided with new knowledge, based on Western science, which "improves, enriches, and changes the content of beliefs."[10]

Progress left destruction in its wake, as traditions fell before the cutting edge of reason. In the present age of "unprecedented change,"

Kawakami recognized that "the need to destroy must also be unprecedented."[11] The Japanese intellectual was stretched on a rack between old beliefs and new learning:

> In every era, a nation cannot avoid ruin if it does not inherit at least some of the thought of its past. But if it does not constantly improve these inherited ideas, continuously accepting the influence of new thought, then not only will it be unable to witness the progress of its culture, but, on the contrary, it will not be able to avoid inviting its own degeneration.[12]

All modern nations shared to a certain extent this need to encourage the scientific learning that produced material progress without losing those irrational elements of traditional life that created a sense of national unity. To delay introducing the latest scientific advances would endanger the nation; but, at the same time, "extreme destruction in the world of thought . . . would overthrow the nation."[13] Kawakami was placed in the confusing position of having to advocate two contradictory policies at the same time:

> This is a time of fierce confrontation between science and beliefs. This is already so in the West and, therefore, it is only inevitable that in our own country, which is rapidly importing a new culture different from our traditional culture, we see a violent collision between science and tradition. To the extent that we do not want to let others win out in the competition among nations, I am asked to devote all my energies to planning the advance and spread of science. On the other hand, I am asked to block with all my might the rapid growth and diffusion of new knowledge in order to prevent disturbances to the national creed . . . Those who most painfully experience this contradiction are probably we who hold a post in the government by dint of scholarship.[14]

In such a situation, where he was forced to "live a life determined by two contradictory orders—to advance and not to advance," Kawakami concluded that the only possible policy was one of gradualism.[15] The task of government and intellectual leadership was somehow to regulate the speed and extent of reform.

> Even though the government should open schools and develop research facilities, the rationalism and knowledge which is the product of such scientific studies must not be allowed to destroy the emotions and beliefs inherited from our ancestors . . . Present-day statesmen must exert their best efforts to reconcile these irreconcilable contradictions.[16]

If Kawakami was particularly sensitive to the strain which Meiji reform goals had imposed on the nation's educated class, it was because, despite the apparent leisurely pace of his academic life style, he always felt driven by a sense of urgency. It was his frantic task as a Westerner to acquire in one lifetime of scholarship not only a professional competence in economic history, German economics, British economics, and finally Marxism and Leninism, as well as the ancillary disciplines of philosophy and sociology, but he had also to discover the cultural assumptions upon which these were based, all the while comparing them with Japanese values. "In the history of economics," he once mused, "I have discovered . . . my own mental history."[17]

Kawakami's significance in the intellectual history of Japan is, however, more than merely as a transmitter of Western thought, for all the while that he was racing to absorb Western thought, he was also trying to fit that thought into his own view of the purpose of learning. That purpose was ultimately moral. It involved a definition of truth in terms of selflessness that seemed alien to Western individualism. He was therefore engaged throughout his academic career in the process of reconciling aspects of Western culture with selected elements of his heritage. Believing that progress came about as a result of technological change, which was studied from the outside, according to empirical methods, Kawakami called himself a scientific materialist. At the same time, he recognized another realm of knowledge which was grasped from the inside, intuitively, and which concerned moral truth. Economics, as a science, belonged to the first category; philosophy to the second.[18] Kawakami's life work involved both. He wanted to be a scientist, but also a sage; a materialist but also a moralist; modern but also Japanese.

The View From Abroad

"If I have a chance to go abroad some day," Kawakami wrote to one of his students, in the spring of 1913, "by all means I would like to study the philosophy of Kant and Hegel." When in that same year the Japanese government satisfied his wish and awarded him a travel grant, he chose to spend his allotted two years in Germany studying not economics but German idealist philosophy. In the same letter he explained the reasons for his choice:

> Bear in mind that there is a limit to how far you can rely on cleverness without having tools. In my opinion there are two kinds of tools. One is the study of methodology itself, the other is the collection of data. The former emphasizes theory; the latter, induction. The former goes deeper into methodology and ultimately reaches the realm of epistomology [*ninshikiron*], while the latter ranges more broadly and ultimately results in the discovery of empirical

laws [*keiken-teki hōsoku*]. Although I do not reject the latter, my natural inclinations prefer the former.[19]

Eager to explore the foundations of knowledge and also to improve his German language skills, Kawakami readily accepted the fellowship offer.

The travel grant was too small to permit him to take his family along, so he left Hide and the children in Kyoto and sailed alone from Kobe, arriving in Brussels in January 1914. After two months of confinement aboard ship, he was more than ready to see some of the cultural attractions of Europe. In February he traveled to Paris, where he lingered for another two months, living in *pensions* and spending almost every day either sightseeing or writing his impressions in essays sent to newspapers back home.

The effect of Kawakami's travels was to make him more sensitive than ever to cultural differences; Europe heightened his patriotic nationalist consciousness. Not long after arriving in Paris, he remarked to friends, "Since coming to the West, we have come to love Japan more."[20] Kawakami was not especially fond of Paris. He noted that, although the city was flourishing owing to the heavy tourist trade, tourism had made Paris decadent.[21] He complained too that the boarding houses lacked intimacy: rooms were dark and meals limited to the continental breakfast of sweet rolls and coffee. Such lodgings made him appreciate all the more the pleasures of the Japanese inn, where meals were brought directly to the guest's room, and the inevitable hot bath in the evening followed by tea provided the traveler with all the comforts of home. In the absence of Japanese amenities and restricted to a modest budget, Kawakami and a friend, Takeda Akira (Sei), another government fellowship holder and law faculty graduate, ate dinner at an inexpensive restaurant called Simonet's. Kawakami's inability to speak a word of French added to his estrangement from French society. In one of his overseas essays he remarked that Europeans lived in public parks and cafes, away from their families. One senses how strange this life style was to him.[22]

Kawakami and Takeda eventually established a small social circle of Japanese friends, including Kawakami's literary hero Shimazaki Tōson, who was staying at a nearby hotel in Paris. He and Takeda violated Japanese etiquette in their eagerness to meet the Japanese novelist. They went to Tōson's room uninvited, but he generously entertained them with green tea and conversation and urged them to return. They visited several times afterward, and each time, the kimono-clad Tōson honored his two guests by serving high-grade green tea in Japanese tea bowls, evoking happy memories of the graciousness of their homeland.

Meeting Shimazaki was one of the highlights of Kawakami's trip.[23]

Conversations in Shimazaki Tōson's hotel room tended to be lively and, upon occasion, heated. Shimazaki remembered one bristling debate between Kawakami and a theoretical physicist named Ishihara on the question of why science had not developed in traditional Japan. Intrigued by the question, Kawakami said it was incumbent upon Japanese university professors to answer it, but he became angry with Ishihara's explanation. Ishihara attributed the lack of scientific tradition in the East to the absence of an interest in investigating the material world and also to the absence of a spirit of practicality. There had been a few mathematicians in old Japan, he admitted, but they had never tried to put their ideas into practice. Warming to his subject, Ishihara further argued that the Japanese always attempted to improve themselves merely by imitating others. They endeavored first to equal China and now they were trying to imitate the West. The trouble was that the Japanese were uncritical in their acceptance of foreign culture, and they never went beyond mere imitation; they were, for example, still putting up with the inconvenience of Chinese characters.

Hearing the soft-spoken Ishihara's critical evaluation of his countrymen as merely borrowers, Kawakami felt obligated to rise to Japan's defense. Agitated, he countered,

> No matter what country, all civilizations have begun by imitating others. The fact that Western civilization is based on Graeco-Roman culture is no different from our ancestors receiving their heritage from China. Not only that, but we Japanese are not necessarily imitators. Look at how many sacrifices our ancestors made when they were importing Chinese culture. Before you know it, we make foreign culture our own. And we improve upon it. The development of Buddhism in our country is one example. Buddhism in India and China is dead, but in our country it's still vital. It's still a short while since we've imported Western culture. Wait forty or fifty years.

Ishihara was not completely convinced by Kawakami's argument. "I'm not quite confident about Japan's present situation," he told him. "We have to do better." "Look fifty years from now," Kawakami retorted. "In the end, Japan's going to do it."

Ishihara had evidently touched a sore spot. The necessity of retaining cultural pride as well as cultural distinctiveness was something Kawakami had been worrying about for some time. It was more than a problem of catching up with the West by becoming completely Western. The total Westernization of Japanese culture, including the Japanese psyche, was, for someone like Natsume Sōseki, the Meiji novelist, impossible: he suffered a nervous breakdown in England trying to write English literature better than the English. The goal of equality was, in addition, undesirable, as Kawakami now was arguing, for reasons of identity: "We can't be satisfied with merely trying to reach [the level of] Western civilization," he said. "We simply cannot become equal in that way. We must think that Japan has her own indigenous and excellent civilization totally different from European culture. Otherwise, we will lose our grounding [*tachiba*]."[24] Japan would have to become equal with the West on her own terms, but not completely the same.

This insistence on the need to preserve and to appreciate Japan's cultural distinctiveness was reiterated in the long series of articles Kawakami sent from Europe to the Osaka and Tokyo *Asahi* newspapers. His impressions of Europe and Japan, reprinted in 1915 as a book entitled *Reflections on Our Homeland (Sokoku o kaerimite),* capture the special pleadings of a patriot abroad as he championed the value of his country's unique characteristics.

The cultural disparity between Japan and Europe had become vividly apparent to Kawakami after only a few days in London. Forced to abandon his studies at the end of his second month in Berlin, when hostilities broke out between Germany and Russia, he had hastily arranged passage to England. Leaving behind most of his luggage and carrying only one suitcase, he fled Berlin on the night of August 15. The display of nationalist emotion witnessed on that night remained fixed in his memory ever afterward: "Parades passed endlessly. The leaders of these parades carried the national flags of the Tripartite Alliance: Germany, Austria-Hungary, and Italy . . . The Germans lifted their voices in unison, crying "Deutschland über alles" . . . hailing each other, waving flags, waving hats . . . Everyone was wild with joy. I've never seen anything like it."[25]

Reaching London three days later, with visions of parades and shouting crowds still floating before his eyes, he was hardly prepared for the response of that city's residents to the war. They were perfectly calm. Nowhere was there any sign of the enthusiasm the Germans had shown. The striking contrast in national character soon set Kawakami thinking further about differences in nationality, the reasons for these differences, and their implications for Japan.

English culture was especially difficult to fathom, German culture less so, but both were patently foreign to him, in spite of his many years of exposure to the West through books and school courses and his ability to read both foreign languages. England's culture was confusing, he wrote, because it had grown like its streets, helter-skelter, with little formal planning. English customs, too, were bewildering to the foreigner. Terms such as Chancellor of the Exchequer or Keeper of the

Privy Seal were difficult to render into Japanese, because they had long ago lost their original meanings.[26]

England was not only confusing, it was disillusioning. In Japanese eyes, England was the most advanced nation in the world, the birthplace of industrialism, of classical economics, and of modern imperialism. But to Kawakami, England was a great disappointment. He disliked the subways and felt oppressed by the crowded London streets. Looking up at the tall buildings in London's financial center, he despaired of ever understanding the capitalist economic system that had made England among the wealthiest countries in the world. He found a university economics lecture dull, and the British Parliament similarly unimpressive; the slow deliberations at the Commons reminded him of a Kyoto University faculty meeting. He attended a talk by George Bernard Shaw on social problems but could not understand Shaw's English, and besides, he had heard somebody say that Shaw failed to practice what he preached. Married to a wealthy woman, he drove around in one of the most expensive automobiles in London. Looking at Shaw's wife seated on the stage, Kawakami cynically decided that Shaw must have married her for her wealth, because he surely could not have married her for her beauty.[27] One of the few things Kawakami enjoyed in London was the vaudeville show.

Critical of English cultural life, Kawakami at the same time exhibited a certain self-conscious defensiveness about Japan's position on the world stage. He seemed bent on proving that Japan could become the equal of the European powers, even though Japan lagged behind the West in material achievements. The Japanese had been endowed with the seeds of greatness, he wrote, and he hoped that Japan would ultimately become the greatest power in the world, just as "we Japanese alone have become the number one power in Asia."[28]

To explain the reasons for Japan's comparatively slower rate of economic development, Kawakami argued that Japanese creative impulses were different: the West was materialistic, Japan was not. If the West's cultural genius was expressed in rationalism, machines, science, and great architectural monuments, Japanese creative power lay elsewhere, in Zen, *haiku,* and the tea house. The Japanese were not actually "late"; they walked a separate path: "The fact that the materialistic culture of the Japanese . . . is backward is a natural result of differences in national character and is not necessarily something to be deeply pessimistic about."[29]

Yet logic and pride compelled the Japanese patriot to ask why Europe had served as Japan's teachers in the modern world. Why had Europeans found nothing worth learning from Japan? Again, the answer lay in cultural differences. Western culture was analytical and therefore easy to understand. Berlin was a good example of this penchant for logical organization. Everything from streets and transportation systems to libraries and parks was rationally laid out and therefore easy to follow. Signs everywhere directed citizens where to go and what not to do. The rational organization of German society had facilitated cultural borrowing. It was not that the Japanese were good imitators, but rather that Western culture was easy to imitate.[30]

Japanese culture, on the other hand, was difficult to assimilate and for this reason the West had not yet discovered its good features. The Japanese appreciated subtlety and vagueness; they liked curved lines and wordless communication. If Westerners analyzed things into their smallest, simplest units, the Japanese bunched things together. His own countrymen would have to plan for the reconciliation of these two cultures, Kawakami wrote, because only they could appreciate both.[31]

This analysis reveals the need Kawakami felt to justify the nature and speed of his own country's development, even if such justification invoked strained comparisons, stereotypes, and false generalizations. He clearly recognized, for example, that English and German cultures were in many ways quite opposite, but at times he lumped both together as the West. His crude dichotomy between the materialism of the West and the spiritualism of the East accurately expressed, if nothing else, his own divided mind.

Possibly Kawakami was forced into this defensive posture by the intellectual currents to which he was exposed in England. One book he read in England was the *Foundations of the Nineteenth-Century,* in which the author, Chamberlain, discussed the relation between race and culture. Kawakami was evidently taken with Social Darwinist theories which attempted to locate the reasons for the West's presumed superiority in race and blood. England held the highest rank among the civilizations of Europe, he explained, because the English had at first mixed many racial stocks, but later had developed a "pure mixture" as an isolated island country. English history offered obvious parallels with Japan, whose people, Kawakami argued, also had at first interbred but then preserved a purity of blood over a period of the last two thousand years. The special history of Japanese blood might explain why the Japanese, of all the East Asian peoples, had become the major power in Asia, even though they resembled Koreans and Chinese in race and physical appearance. Their blood was the same, but its history and quality was basically different, which was why the Japanese were "exhibiting an amazingly unique culture."[32] Japan thus constituted a middle world which was neither Western nor Asian, but something in-between, unique and special.

Kawakami's brief interlude in Europe stimulated his patriotic pride and his competitive spirit. He did not share the complete disillusionment with Western civilization that characterized the response of Chinese intellectuals who visited Europe several years later. Liang Ch'i-ch'ao, in 1919, was appalled by the carnage of war; writing on the eve of war, Kawakami was excited. Imminent conflict between the great powers did not turn him away from chauvinist statements; rather the war fever intensified his own patriotic emotions. He wrote to his Japanese readers that he was pleased to find Japanese goods on sale in an English village, and when a Japanese performer appeared on the stage of London's Albert Hall, Kawakami, seated in the audience, was so overjoyed, tears spilled down his cheeks.[33]

While living abroad, Kawakami discovered much to value in his own country's culture and, because of this, his *Reflections on Our Homeland* exudes an optimism and even exuberance which for a brief moment he allowed himself to enjoy. Only at the end of this collection of his European essays did he temper his optimistic forecast with more sober observations: he had viewed the cloud of poverty and class struggle hanging over England and saw it spreading to Japan.

Measured in terms of his evolution as a Marxist, the two and a half months Kawakami spent in England prior to returning home were perhaps among the most important in his life. He was not aware of their significance at the time; instead, he felt frustrated and restless in the face of the unexpected events that had left him stranded in England, without his luggage, short of funds, and lacking any clear study purpose. Yet, the profound, though delayed effect of his experiences in England influenced the later course of his academic career.

Loneliness was not his major problem. Throughout his overseas stay, Kawakami was usually in the company of at least two or three of his countrymen. In addition to Takeda, he met a high school teacher named Tozawa Shōhō with whom he shared lodgings in London, and a specialist in criminal law named Tomida. The men often boarded at the same rooming houses, taking meals and traveling together.[34] Several Englishmen befriended him, and he was regularly invited for tea at the home of one acquaintance.[35]

Some of Kawakami's discomfort may have been triggered by his spending habits. He depicted himself as more serious than other overseas Japanese students who tended to squander their money on women; however, he splurged in other ways. Although he often sermonized on the virtue of frugality, Kawakami also frequently admitted, "I am a man who cannot hold on to money."[36] His fellowship was hard to budget, because it was paid in semi-annual allotments. Kawakami was inclined to use the money immediately upon receiving it, and his purchases were often unwise. In Germany he bought a silver watch guaranteed to run for twenty years; the watch soon stopped running. He bought a similar pocket watch in an exclusive shop in London; and although it lived up to its lifetime guarantee, thereby permanently endearing him to the "reliable" English people, the cost of the watch and also of an expensive raincoat which he would wear only once—on the day he returned to Japan—left him without funds to pay for his boat ticket. He had to borrow the money from the Japanese embassy.[37]

Haunted by the sense that he was wasting not only money but precious time, Kawakami yearned for the solitude of scholarly life, convinced as he was that he could learn more from books than travel. For better or for worse, he was "the hermit of Yoshida," the hill behind Kyoto University.[38] Yet, much as Kawakami's unhappy stay in England reinforced his ivory-tower predilections, his personal experiences while abroad actually had a great impact upon his thinking about economics and especially about the forgotten problem of poverty. By reading newspapers, Kawakami acquired a picture of European life that led him to question the real wealth and strength of the West. A letter in the London *Daily News*, for example, complained that the writer could not afford a pair of shoes. An even more vivid encounter with poverty in England came from Kawakami's brief stay in the home of a poor tenant cultivator some seventy miles southwest of London.

Kawakami's own desperate financial predicament had led him to seek cheaper lodgings in the countryside. He traveled by train as far as the region around Ramsey, where he and a friend rented a room from a farm family. The village was dominated by a few aristocratic families, who hired the villagers to work their land. The contrast between the obvious wealth of the country squires and the poverty of their tenant cultivators was all too apparent to Kawakami. He was struck by the extravagant use of land: the boulevards leading from the road to the manor homes were broader than Kyoto's main streets. He was also surprised by the vastly different life styles of these two rural classes: while the one went fox hunting, the other labored long hours for little reward. Given such unequal distribution of wealth, Kawakami wrote, shortly before leaving England, it was no wonder that class struggles emerged: "Just as there are endless wars among human races throughout the world, so within each country we can expect class war to become more and more widespread."[39]

This disturbing prediction was written in November 1914. Several months later, in the spring of 1915, Kawakami reached Japan, where he immediately returned to his desk to begin a study of the problem of poverty in capitalist societies. Although he was not aware of it at the time, he had already taken his first major step on the road to Marxism.

Kawakami Hajime in 1906   in 1924 with his parents, wife, and children

in 1928 at a Labor-Farmer party convention

in 1944

part 2. academic marxist

学門

6. the road to marxism

"It is surprising that a large number of people in the modern civilized world are poor," Kawakami Hajime wrote in 1916, beginning a series of articles that electrified the Japanese intellectual world.[1] The articles, which ran for over three months in the Osaka *Asahi* newspaper, enjoyed an immediate success: one month after the final installment, they were compiled for publication in book form, and between 1917 and 1919, thirty reprintings of *Tale of Poverty* (*Bimbō monogatari*) were made to satisfy public demand.

The *Tale of Poverty* suggests the latent influence of Kawakami's brief overseas tour on his future evolution as a Marxist. While that influence was minimized by Kawakami himself (there was nothing to learn in Europe, he told his student, Kushida Tamizō)[2] and has been almost completely overlooked by Japanese scholars, nevertheless, it was Kawakami's first-hand observations of social problems in wartime Europe and especially in England that inspired his famous series of articles. Like so many Japanese before him, Kawakami had gone to Europe expecting to be dazzled by the opulence of the world's most civilized nations; he may have even hoped to discover some new aspect of Western civilization that would account for the West's enormous power, or to glean the latest developments in Western thought that could be pressed into service back home. Instead, what had begun as a search for the secret of the West's strength, had turned into a probing analysis of the reasons for its weakness.

The Tale of Poverty

Kawakami's experiences in Europe caused him once again to reevaluate the capitalist economic system, because he had discovered that, "even though England, the United States, Germany, and France are exceedingly rich, their people are exceedingly poor." This simple fact was surprising, because the Japanese had associated poverty with underdevelopment and had assumed that, by simply increasing industrial productivity, Japan could eventually equal the wealth of the European powers. It was therefore "somehow odd that in the various enlightened countries of the West, where the use of machines is the most widespread, an extremely large number of people are poor."[3]

The implications of this discovery were far-reaching. Poverty was not just the blight of underdeveloped countries; it was also found in the whole capitalist system. Kawakami had earlier questioned the premise of classical economics, wondering how, despite its hedonistic and individualistic principles, the capitalist system worked for the good of the whole society. The answer he gave now was simply that it did not. He came to the conclusion that, "the reason why . . . [England, Germany, and France] are still called the wealthy nations of the world, despite the fact that they have many poor people, is that an amazing fortune has been concentrated . . . in the hands of an extremely small number of people."[4]

Drawing on data from studies of the poor done in England and the United States, Kawakami marshaled solid evidence to prove his contention that widespread poverty existed in the major countries of the Western world. Charts, graphs, and tables quantitatively illustrated inequalities in income distribution. Information compiled in Charles Booth's *Life and Labour of the People of London* (1902); Charles Bowley's *Livelihood and Poverty* (1903); King's *The Wealth and Income of the People of the United States* (1915), and numerous other English-language materials, lent additional support to his contention that poverty had become a serious problem for the nations of the Western world. Indeed, Kawakami advised his readers that in the *Theory of Social Revolutions* (1913), Brooks Adams had predicted an outbreak of major revolutions before 1930.

In the introduction to the Iwanami edition of his book, Kawakami explained why he was so concerned about the existence of poverty. His explanation demonstrates the enduring Japanese bent of his humanist thought, an element that critics later decried. "I want to eradicate poverty from human society," he wrote, "simply because poverty is a hindrance to hearing the Way."[5] He wished to distinguish his position from those who were interested in material wealth for its own sake. Too many economists defined the goal of life in purely materialistic terms. This was a fallacy Kawakami had fought throughout his adult life. The exclusive concern with material goals that, in his eyes, characterized both capitalist and socialist economics had made him hesitate fully to accept either one: "Even though a group of economists measures civilization only in terms of the progress of material culture—the increase of wealth—I believe that cultural progress in its true meaning lies only in having as many people as possible hear the Way."[6]

Kawakami's purpose in writing his series of articles on the controversial question of social problems was not merely to publicize the existence of poverty, but more important, to analyze the reasons for economic inequities and to propose cures. Not surprisingly, he approached these questions from the viewpoint of a moralist as well as a social scientist.

Kawakami was now prepared to reject the explanation of the origins of poverty offered by Malthus, whose gloomy predictions for society's future had stirred young Hajime, some fourteen years earlier, to write a melancholy letter to Katayama Sen, confessing his doubts about economics as a career. According to Malthus, the tendency for population to increase at a faster rate than the food supply made poverty inevitable. But this theory, Kawakami argued, did not explain why only some people were poor, while others were rich. Further, with the advent of machine technology, it had become possible for material productivity to outstrip population increase.[7] Hence poverty could not be attributed to overpopulation. Rather, the root of the problem lay in economic individualism. "The unshackled activity of the self-interest motive at the bottom of today's economic system actually foments a lamentably unhealthy situation."[8]

In present-day society, Kawakami argued, money had become more important than anything else: people fought over it, and they worshiped it, losing all capacity for human sympathy. The great aim of economics, according to Adam Smith's *Wealth of Nations,* was endlessly to increase the wealth and power of a nation, but Smith had overlooked an important truth: "Wealth is essentially nothing but the means for a man to become truly a man—the aim of human life. The amount deemed necessary to achieve such a goal is not unlimited."[9]

In Kawakami's opinion, Adam Smith had made a fatal error in establishing economics as an independent science outside the realm of ethics. It was an ironic fact that Smith, who had begun his career as a professor of moral philosophy, was responsible for affirming economic activity on the basis of self-interest, thereby setting the study of economics apart from moral considerations. Such an individualistic economic theory relieved society of responsibility for the material well-being of its members. The disastrous results of economic competition uninformed by moral restraints were readily apparent in England where, in order to ease domestic problems, leaders had found it necessary to go to war "for the sake of competition in the export of capital":

> If the money bags and the capitalists of England would perceive their real responsibilities as consumers and as producers, then not only would it be possible to solve peacefully social problems within the country, but also it would be possible to maintain peace in the world, for according to *The Great Learning,* the man who would make the light of virtue clear to the world, must first set his own house in order.[10]

These ethical observations led Kawakami to the central part of his analysis. Assuming, as he always had, a direct connection between the hedonistic spending habits of the rich and the misery of the poor, he concluded that the real cause of poverty was the excessive demand of wealthy people for luxury goods. Rich consumers had diverted society's productive forces to frivolous ends. As a result, luxury items abounded, whereas daily necessities were scarce. As long as the rich could afford to make these extravagant purchases, a demand for them would exist, and producers would respond accordingly by manufacturing more. The key point, Kawakami emphasized, was that demand did not necessarily mean need but only available money to spend. A small proportion of the population was creating the demand that stimulated increased production. This fact explained why, despite rising rates of production, a nation could have so many poor people.[11]

Kawakami's proposed cure for the gaps between rich and poor followed logically from his diagnosis of the cause. He suggested that the rich should refrain from buying luxury goods, so that manufacturers would be forced by the natural laws of economics to produce necessities for the poor instead. It was not necessary, he pointed out, to drive automobiles or to build tall buildings; nor was it necessary for people to eat as much as they did. By writing on both sides of a sheet of paper and watering trees with used water, further economies would be possible.[12] Wealthy consumers should practice frugality and producers should act like public servants by managing their business enterprises as if they were "making a profit for everyone":

> Accumulating a fortune is not a bad thing . . . only, since one always uses it wastefully, when it passes to another person's hands, I think if you were to make a lot of money under the belief that you, yourself, were managing that wealth, for the sake of society, then it would be perfectly justifiable. No matter how much money they make, and how much of a fortune they accumulate, people who share such a faith can be expected not to use their wealth on luxury expenditures for themselves and their families. In this way, all social problems will be satisfactorily solved and for the first time, there will be a reconciliation between private enterprise and ethics, an accord between economics and morality.[13]

The solution to the problem of poverty, in other words, was to increase production while practicing frugality. Such moral panaceas, however, were similar to Kawakami's early writings and scarcely differed from the suggestions of Tokugawa thinkers writing one hundred years before. What Kawakami had done was to blend a modern analysis of economic problems in terms of supply and demand with a traditional cure—moral reform.

In addition to voluntary constraints on consumption, Kawakami offered two other possible cures for poverty. One was the nationalization of industry and the other was government measures to redistribute wealth. Germany and England respectively provided striking contemporary examples of each.[14] In his study of English welfare legislation and programs, Kawakami had discovered the numerous inventive uses of social policy in curing social problems. The British school lunch program and old-age pension systems were only two examples of the types of social welfare legislation championed by Lloyd George and passed, against the protest of the rich, in a country that had spawned the original doctrine of laissez-faire economics. Nationalization of industry—the other approach to curing poverty—was being tried in Germany, where the government in 1915 had established control over bread and bread products.[15]

Although some scholars viewed these policies as socialistic, Kawakami preferred to see them as the enactment of nationalism.

> Socialism has many different meanings and no fixed definition . . . but it does not recognize the existence of the nation as the basis of economic organization; that is, it upholds internationalism, emphasizing the profit of the working classes, and it is readily confused with anarchism. Therefore, we avoid the word socialism and say nationalism, even though it can be argued that the nation is just one form of society. By [nationalism] we mean fusionism [*gōdōshugi*] or government enterprise, as opposed to individualism or private enterprise.[16]

Whatever the terminology, it was clear to Kawakami that the "economics borne of Adam Smith has already completed its mission and now is truly the time when a new economics is to be born."[17]

This confidence in the victory of morality over the ethos of capitalism was further encouraged by certain European critics of industrial society who, Kawakami learned, shared his belief that the capitalist system, if chastened, could serve the spiritual as well as material needs of mankind. They, too, believed that the motives underlying economic behavior could be altered, without necessarily changing the institutions of capitalism. The following quotation from Thomas Smart, printed in English in an article Kawakami wrote in 1917, is representative of the ideas of this humanist school, to which Thomas Carlyle and James Ruskin also belonged:

> My thesis . . . is that, as the only business which mankind is interested in preserving is, fundamentally, the service of man, this bread-and-butter life may be taken up by all from highest to lowest, and yet be transformed into a life of the loftiest moral purpose by consciously adopting it as the Service of Man . . . the present system is

worth saving—not only for what it has done, but for the fact that, guided by conscious moral purpose, it may be reconstructed to serve still higher ends. It is not a reconstruction of the economic life, but a reconstruction of its Motive.[18]

Drawn to these British critics, whose moral and esthetic revolt against capitalism confirmed his own ethical inclinations, Kawakami overestimated their importance in the world of European economics. His assessment of them as progressive leaders in the study of economics was misleading. Neither Carlyle nor Ruskin was really an economist. Both were romantic and even reactionary figures in their times, representatives of a mid-nineteenth century "Victorian underworld of economics" with less influence on their own society than Kawakami assumed. In Kawakami's interpretation, however, they emerged as founders of a new economics that breathed life into the rigid system of economic "iron laws" the classical economists depicted.

The important message in Kawakami's *Tale* was that, one way or another, Europeans were already moving to replace the heartless, money-grubbing aspects of the private enterprise system with what Kawakami called a humanist economics. For young Japanese intellectuals, attuned to the latest in technological and intellectual developments from abroad, this was exciting news. It took them a number of years to conclude that, in Ōuchi Hyōe's words, Kawakami had merely "shown us a new side of western Europe's bourgeois economics linked to Japan's old morality."[19] But in the meantime, by spawning a movement to establish this humanist economics, Kawakami helped lay a major steppingstone on the road to Marxism in Japan.

Reading the book today, and recognizing the familiar Confucian-sounding bromides that guided Kawakami's approach to economics, it seems difficult at first to explain why *Tale of Poverty* had the enormous impact it did on the younger generation of post-World War I Japanese. Only when we place it against the background of its times can we appreciate why this little volume of less than two hundred pages in the Iwanami paperback edition (including a picture of Adam Smith in the frontispiece) catapulted Kawakami once again into the limelight, this time as a leader of the new humanist movement.

Tale of Poverty was one of the first published works to dissolve the conspiracy of silence that had stifled critical thinkers since the end of the Meiji period. Removing the problem of poverty from the scholars' conference table and placing it before the general reading public, Kawakami's book helped instill in the younger generation of Japanese students a new sense of social responsibility, much as Kinoshita Naoe and Tanaka Shōzō had inspired him in his own student days. Ōuchi Hyōe recalls how his generation was "told of the existence of poverty in Japan and moved by Kawakami's explanation of it."[20]

A further effect of Kawakami's book was to stimulate an interest in the study of economics, an academic area still little known to younger students. In his preface, he called the "economic problem one aspect of the human problem," encouraging students to believe that they could apply their knowledge of economics to the solution of poverty. "From *Tale of Poverty*," wrote Sakisaka Itsurō, the Marxist economist, "I learned that 'poverty' was not only my own problem, but the problem of today's society, so I knew that this problem of poverty was something worth devoting one's entire life to, just as many intellectuals and scholars had done. This was a great discovery to me."[21] The original and distinctive element in Kawakami's otherwise outmoded theory of poverty, in other words, was the link he suggested between the study of modern economics and the cure of social problems.

Tale of Poverty was probably Kawakami's most important professional work; considering that he was primarily a translator and interpreter of Western texts, it was also his most original work. Its historical value for readers interested in Kawakami Hajime's thought lies in the way it captured the many conflicting influences impinging on his consciousness: in his undifferentiated treatment of socialism and nationalism, his blend of European humanism with traditional moral teachings, and his combination of classical economics with Confucian admonitions, Kawakami held together for one final moment the numerous intellectual currents that he had for so long struggled to integrate into one system of thought. All the themes of his mental life—curing poverty, hearing the Way, serving the nation, and reconciling science and religion—were joined in his loosely woven analysis. The finished product struck his many avid readers at the time as a ground-breaking piece of scholarship that placed economic theory in the service of human welfare.

However conventional his moral recipe for reform, by exposing at the height of the wartime industrial boom in Japan the fallacy of the government's preoccupation with productivity alone as a measure of national wealth, Kawakami took a dramatic step away from social policy and toward socialism. Focusing on the poverty of industrial societies rather than the wealth of nations, he set the stage for his eventual shift in loyalties from nation-state to society.

The Influence of Kushida Tamizō

At the time of its publication, Kawakami considered *Tale of Poverty* to be his major professional accomplishment in over a decade of research. "I have written many things in the past ten years," he said in the preface, "but nothing can equal this work."[22] We know from personal accounts that many students among the more than one hundred thousand readers of the *Asahi* at that time were inspired to study economics after reading Kawakami's articles and flocked to Kyoto in the following years to study under him. Sakisaka Itsurō recalled that when he read the *Tale of Poverty* in his second year of high school, he found it "so absorbing I could hardly wait for the newspaper to arrive."[23]

Almost immediately after its appearance, however, the book came under attack from economists of a Marxist persuasion. Shaken by their criticism, Kawakami abruptly stopped publication in 1919, resolving to "wash my hands of bourgeois economics and prepare to study Marxist economics."[24] *Tale of Poverty* thus represents the major turning point in Kawakami's path to Marxism. His "greatest work to date," the book that made him rich and that assured him a permanent place in the history of Japanese economic thought, *Tale of Poverty* was also the crushing failure that eventually led him to abandon social policy in favor of scientific socialism.[25]

Crucial to Kawakami's eventual decision to espouse Marxism was the criticism of his theory of poverty delivered by his former pupil, Kushida Tamizō (1885–1934), who was rapidly assimilating the new theories of Marxist economics that began entering Japan toward the end of World War I. Kushida, only six years younger than Kawakami, had studied under him at Kyoto Imperial University. The son of an impoverished family from Fukushima, Kushida had inherited none of the Meiji *bushidō* spirit that so characterized his teacher's personality and was untroubled by the moral preoccupations that inspired all of Kawakami's work. Although in temperament they were opposites, the emotional and even histrionic Kawakami and the more rational and even-tempered Kushida found they had much to share with each other. The many long hours they spent talking together in Kawakami's home cemented a bond between them strong enough to withstand their intellectual differences.[26]

Kushida was indebted to Kawakami for help in obtaining part-time work translating German to finance his stay at the university.[27] After his graduation in 1912, he again called upon his teacher for assistance, first in finding a job and later, in finding a wife. The two differed as much in their taste in women as in their approach to economics: Kawakami soon learned that his pupil would accept neither his theory of poverty nor his selection of a bride, and in the end, he allowed himself to be beaten down on both issues.[28]

When in 1916 Kushida chose to dispute Kawakami's theories in one of his first published works, he was not a confirmed Marxist, although he was grappling with certain Marxist concepts. He challenged Kawakami's basic explanation of poverty, arguing that the cause of poverty lay not in the consumption of luxuries by the wealthy, as Kawakami had asserted, but in the exploitation of the workers by the capitalists. Consequently, the basic solution of the problem rested not in the reform of individual morality, but in the reform of distribution; that is, ultimately, in the reconstruction of social organization. Instead of urging ethical self-awareness on the part of the wealthy, one should rather seek economic self-awareness on the part of the workers, because the capitalist profit motive could be eliminated only by reconstructing class organization on the basis of the workers' class consciousness.[29]

Kawakami was unconvinced by these arguments, for even while admitting "there were errors in my explanation,"[30] he had begun shortly afterward to write *Tale of Poverty*. Once again, and more forcefully, Kushida had criticized his former teacher's position. While respecting Kawakami's "sincere attitude," he had found his thinking "outdated," for in *Tale of Poverty*, Kawakami had tried again to combine science and ethics.[31]

Under the pressure of Kushida's persuasive arguments, Kawakami began to revise his own thinking, taking an interest for the first time in political revolution as a mechanism for progressive change. "Authority is the instrument of aristocracies, revolution is the engine of democracies," he wrote to Kushida, three months before the engine of democracy was set rumbling through Russia. Quoting in English from Delbert's *Social Evolution* (1902), he continued,

> Aristocracy and democracy are the flow and ebb of civilization, and their currents may be clearly defined. One acts as an element of organization or of creation, the other as an element of destruction or of variability; otherwise evolution would be inexplainable. The sentiment of duty preponderates in aristocracies, with democracies that of their rights preponderates.[32]

He even began to see an affinity between revolution and his own youthful idealism. After reading Thomas Godwin's *Political Justice,* he scrawled on the back of a postcard to Kushida, "How much Godwin's *Political Justice* fits my Selfless Love period. When I think that he wrote it in the very middle of the French Revolution, my interest is more than a little. The French Revolution is really an interesting and important chronicle of the course of human life, isn't it?"[33] Nevertheless, faith in the rejuvenating powers of moral reform and in the efficacy of the social policy propounded by bureaucrats, Imperial University professors, and other contemporary economists of the day had sustained Kawakami for many years in his resistance to socialist theories, and he hesitated to accept Kushida's criticism without long and careful deliberation.

Kawakami was not completely unfamiliar with socialist thought. He had long harbored mixed feelings of attraction and repulsion toward the socialist creed, applauding, in his youth, the goals of the Christian socialists but questioning whether these goals were attainable. He knew that socialism had been transformed significantly at the hands of the Marxists and suspected that its scientific basis made it different from his earlier understanding of it. For this reason he had voiced opposition in 1912 to the ban on socialist thought that followed the Kōtoku Shūsui affair, pointing to the growing popularity of socialist thought in Europe as proof that it deserved a hearing in Japan. The pressures of his own professional commitments combined with the scarcity of materials on Marxism, however, had prevented him from keeping abreast of these latest developments, making him "dull-witted," he wrote ruefully in his *Autobiography*, in his efforts to understand Marxist theory.

Kawakami was by no means alone in this painful period of re-evaluation which followed the publication of *Tale of Poverty*. The issue of socialism dominated meetings of the Association for the Study of Social Policy, where debates over the differences between social policy and socialism, reminiscent of similar debates raised in the early days of the association's founding, reflected the great schism developing among intellectual leaders in and out of the academic world. Kawakami, Kushida, and Fukuda Tokuzō (1874–1930), a professor of economics at Tokyo Higher Trade School and later an editor of *Liberation (Kaihō)* magazine, were some of the speakers at the annual meetings of the association who debated the relative merits of social policy and socialism in increasingly more strident tones. These internal disagreements, combined with the rapid diffusion of Marxist thought and the growing labor movement, undermined the organization's usefulness and, as questions of social reform rapidly spread from bureaucratic and academic elite circles to the popular press, the labor union cell, and the university campus, the Association for the Study of Social Policy declined in importance and was finally disbanded in 1924.

Kushida's criticism of Kawakami's theories typified the conflict between advocates of social policy and proponents of socialism in these years during and immediately following World War I. In terms of method of reform, Kushida's criticism caught Kawakami between moral reconstruction and political action, between reconciliation on the one hand, and class struggle on the other. On the level of economic philosophies, Kushida's argument trapped Kawakami and other spokesmen for social policy between the economics of production (capitalism) and the economics of distribution (socialism). The first presumably served national interests; the second solved social problems. With few voices heard in defense of a middle ground between these two ideologies, Kawakami was faced with an either/or decision. As long as he had treated poverty in terms of insufficient production, he had been able to forestall such a clear-cut choice. He soon realized, however, that:

> If the aim of economics is simply the greatest production of wealth, that basic problem of economic policy was already solved about 150 years ago by Adam Smith. But there are really two basic problems of economic policy: the first is the problem of production; the second, the problem of distribution. Thus, the most basic and unresolved problem in the present-day world of economic studies is how we should reconcile these two ... for all modern economists, this problem still requires investigation.[34]

The decision to study Marxist economics was not an easy one to make, because it involved an act of personal and political commitment whose implications were dangerous and far-reaching, and not simply the choice of a new subject of academic pursuit. True, Kawakami could justify Marxist studies in academic terms as research and, in so doing, postpone grappling with the thorny political choices that would eventually confront him. Then, too, journalism provided him with a halfway house between the world of thought and the world of political action, though he knew the risks involved merely in writing about Marxism at a time when the word socialism or even sociology was suspect.

It was no wonder that he now recalled his experience in the Selfless Love movement some twelve years earlier, for once again fate was daring him to drop out of the charmed circle of Imperial University professors and become instead a "social repairman." But the stakes were higher in 1917 than they had been in 1905. Approaching middle age, he could not leave his family, resign from his teaching position, and plunge into social activism with the same reckless disregard for consequences that had allowed him to leap into the Garden of Selflessness. His son had been ailing for many years, his teaching responsibilities were onerous, and he was in the midst of delicate negotiations over the establishment of a separate faculty of economics at Kyoto Imperial University, a goal that meant a lot to him and occupied a great deal of his time. It was disheartening to learn that, having finally arrived at a preeminent position in society, his professional expertise was being called into question. Confronted with the prospect of assimilating a whole new body of economic thought, for which he lacked both training in economic theory and the time to acquire it, he understandably hesitated to switch intellectual allegiances. Depressed by the decision confronting him so late in his career, Kawakami depicted himself as a prisoner waiting sentence: "Sitting here in prison...I greet the spring of my fortieth year."[35]

Nevertheless, it was not so much personal considerations as the specific areas of disagreement between himself and other economists that left him uncertain in the years between the publication of *Tale of Poverty* in 1917 and his decision to embrace Marxism in 1919. At stake was a view of government and morality that had become second nature to Kawakami; he was born and raised with the ethos of *shishi,* an elitist view of statecraft that left reform in the hands of the qualified few to rule on behalf of all. His entire political philosophy rested upon the belief in an identity of interests between government and society mediated by moral statesmen. Such a view was implicit in his approach to economic reform through social policy. In the name of social harmony, he had pressed the virtues of frugality on rich and poor alike, always striving to reconcile conflicting class interests through ethical restraints. He was now asked to vilify the bureaucrats with whom he had previously discussed solutions to economic problems and to place his faith in the poor, whom he had always worried about, but never personally known. It was one thing to suggest and even demand economic relief for the poor, but another thing entirely to agitate among them, raise their political consciousness, and work to place power in their hands. After all, what were their qualifications for rule? Only poverty, not education, or moral cultivation, or dedication to the common good. Alleviation of their grievances was the goal of good government, but did this justify empowering men who were exclusively concerned with their own material well-being? Events in Japan after 1917 seemed to hold the answer to that question.

The Russian Revolution and the Japanese Rice Riots

Although several years passed before detailed information about the Russian Revolution of 1917 reached Japan, the impact of the Revolution on Japanese political life was felt almost immediately. The collapse of both the czarist regime and, soon afterward, the German kaiser's government seemed to herald a new age of popular government. Such an interpretation was seen in the liberal Osaka *Asahi* newspaper's interpretation of the Revolution as a "victory for popular democratic forces over bureaucratic government."[36] For critical Japanese intellectuals, who likened Japan's bureaucratic government to the oppressive Russian autocracy, the Russian Revolution, together with the triumph of the democratic nations in World War I, confirmed their view that the age of democracy had arrived.

One immediate effect of the Russian Revolution on Japan was to encourage reformers from many different ideological camps to enter the political arena and agitate for popular causes. Almost overnight a number of new journals appeared dedicated to introducing socialist thought, and the labor movement, which, under Suzuki Bunji's guidance, had grown by 1918 into an organized drive of some 107 unions, now demanded legal recognition and called for universal suffrage for men. While the Revolution did not immediately influence the conduct of domestic Japanese politics, it did lend encouragement to the workers' movement by providing evidence that the trend of the times sanctioned their efforts to achieve greater political power.[37]

These optimistic predictions for change through popular foment naturally influenced Kawakami's own views on reform methods. His entire moral-economic scheme had rested upon an identity of interests between state and society. This bond between the two now began to dissolve as he gradually redirected his mental energies from the cause of the nation to the people within the nation. The dictates of absolute unselfishness which had led him to embrace national interests were slowly transferred to social problems. Whereas earlier he had relied on social policy to reconcile the conflict between workers and capitalists in the interests of national industry, now, in 1918, he declared that "social policy is nothing but production policy."[38] In the light of his former allegiances, that simple sentence was not merely a statement of fact; it was a denunciation. By the spring of 1918 he was prepared to concede in deference to Kushida that it was necessary "to rebuild the system through class warfare, based on the economic self-awareness of the workers."[39]

Nevertheless, between concession to his critic and total acquiescence in his theories lay an agonizing period of intellectual reappraisal. Evidence of his indecision is the fact that he continued to publish articles on Carlyle and Ruskin, the romantic critics of capitalism whose moral sentiments approximated his own. Moreover, he included all of *Tale of Poverty* as well as articles on Ruskin and on social policy in *Views on Social Problems (Shakai mondai kanken)*. Published in the fall of 1918 the book held together in tenuous balance Kawakami's long-cherished moral and economic ideas and his doubts about these ideas.

Events in the latter half of 1918 tipped the scales of Kawakami's precarious balance between capitalist and socialist economics. Midsummer riots in the cities, sparked by an inflationary rise in rice prices, provided impressive proof of the seriousness of the economic problems besetting the nation. The immediate cause of the rise in prices was the Japanese military intervention in Siberia, but inflation was symptomatic of the more general phenomenon of a rapidly increasing urban population, drawn to the cities by the quickened industrial pace accompanying Japan's participation in World War I. In a manner typical of

Kawakami's moral approach, he urged the government not to suppress the riots with violence but to give the people ideals and hope.[40] The Terauchi cabinet responded instead by trying to censor the Osaka *Asahi* newspaper's coverage of the riots. When one reporter referred allusively to a rainbow penetrating the sun, a Chinese omen for impending dynastic collapse, he was arrested. In protest against the government's interference with the press, the liberal faction of the newspaper, under its editor-in-chief, Torii Sosen, resigned, and Kawakami, whose "The Problem of Rice Prices" had appeared in a series of articles written for the Tokyo and Osaka *Asahi* newspapers from August 18 to August 24, also discontinued his association with the paper.[41]

The Osaka *Asahi* affair was the last major landmark on Kawakami's road to Marxism. Soon after this incident, he was drawn into the burgeoning workers' movement. Members of the Kyoto branch of the Friendly Society and the Worker-Student Society (*Rōgakkai*) were instrumental in gaining Kawakami's cooperation in their drive to legalize labor unions. The Kyoto Friendly Society, formed in February 1917, was headed by Takayama Gi'ichi, a student at the law faculty of Kyoto Imperial University.[42] Suzuki Bunji, head of the national federation of Friendly Societies, helped Takayama establish the Kyoto branch, and on September 17, 1917, Suzuki met over dinner with Kawakami and other professors to discuss workers' problems.[43]

A year later the Worker-Student Society was formed, also at the initiative of Kyoto Imperial University law students. Galvanized into action by the rice riots in the summer of 1918, they organized their group for the purpose of studying socialism and labor problems.[44] Takayama was the founder of this group, too, and among the other members were young students like Horie Muraichi, Shiraishi Bon, and Kobayashi Terutsugi, who had all come to Kyoto to study under Kawakami.[45] Almost from the start, Kawakami became the real leader of the group.[46] His *Tale of Poverty* and his series of articles on the rice riots made him the natural spokesman for their concerns and his university position lent the necessary academic respectability to their campus meetings. "Although we were under the rule of military authorities and the police," Kawakami later wrote, "in one corner of the university we could preserve our freedom of speech and thought."[47]

Even before the autumn of 1918, when the Worker-Student Society was founded, Kawakami had come to support workers' demands for freedom of speech and thought, the right to organize and to strike, and the power of the ballot. After the summer rice riots, he became further convinced of something he had already grudgingly admitted to Kushida—the necessity of class warfare. His increasingly more radical position did not completely satisfy his student following, however; and

Takayama Gi'ichi, the spearhead of the student-labor alliance, even chided Kawakami for some of his public statements, saying, "Anybody can be brave in the middle of an ivory tower."

The need to demonstrate his sincerity to progressive students may explain why one of Kawakami's earliest speeches to the Friendly Society was more radical than members were accustomed to hearing even from their own leader, Suzuki Bunji, a Christian convert who espoused conciliation and compromise between labor and capital. Speaking in December 1918 before a meeting of the Federated Friendly Societies in Kyoto, Kawakami suggested that class war was inevitable: "The working class and the capitalist class, being two classes which stand in opposition to each other, cannot be reconciled." Kawakami's pronouncement won the approval of Takayama's followers in the Worker-Student Society and the Kyoto Friendly Society, but it alarmed other members of the audience, who immediately sent a telegram to Suzuki in Tokyo. Shocked, Suzuki hurried to Kyoto to see Kawakami and the following day, he conferred with Kushida, who was teaching at nearby Dōshisha University. While we do not know the exact substance of these discussions, Suzuki afterward assured Friendly Society members that there was no difference of opinion between himself and Kawakami or Takayama. This incident became a turning point for the Kyoto branch of the Friendly Society, which gradually moved toward socialism after this time. Takayama proclaimed, "Dr. Kawakami's thesis is the aim of the Friendly Society."[48]

Shortly afterward, in an article published on New Year's Day 1919, Kawakami alluded in metaphor to his new and more radical views. His "Soliloquy of an Unnamed Doctor" ("Aru isha no hitorigoto") recommended a "surgical operation" to remove parasites housed in a patient's intestines. To cure the patient, "a little blood must flow," he warned, but there was no other way to restore the patient to good health. Kawakami, the physician for Japanese society, was prescribing drastic reform and possibly even revolution.[49]

By the time he wrote this article, Kawakami had already halted publication of *Tale of Poverty* and resolved to "utilize to the greatest extent possible my position as a university professor ... [in order to] propagate socialism,"[50] but having severed his connections with the Osaka *Asahi* newspaper, he felt he had "lost the stage from which to speak to the people."[51] Beginning with the *Tale of Poverty* in 1916, Kawakami had been regularly publishing articles in the *Asahi* newspaper. In September 1917 the paper had published his series of articles on "Marx's *Capital*" and in 1918 it had carried his articles on the rice riots. Deprived of this public forum, Kawakami now cast about for a new channel of expression.[52]

Other writers formerly associated with the *Asahi* newspaper were, in the meantime, also seeking new publishing outlets. Maruyama Kanji (father of Maruyama Masao) founded a newspaper, the *Taishō Daily (Taishō nichinichi)*, and in February 1919 Hasegawa Nyozekan established the scholarly magazine *We (Warera)*. Ahead of these two other independent enterprises, Kawakami decided to put out his own private journal, calling it *Research in Social Problems (Shakai mondai kenkyū)*. Two of his friends, Kushida Tamizō and Kojima Yūma, a faculty member at Kyoto Imperial University, arranged to have Kōbundō Publishing Company publish it on a monthly basis.[53] The first ten issues of the journal, starting in January 1919, were devoted to an exposition of Marx's historical materialism. This marked the beginning of Kawakami Hajime's serious study of Marxism.

7. the meaning of marxism

Kawakami Hajime was barely familiar with Marxism when in 1919 he announced his decision to propagate it. Although he quickly succeeded in his pledge "to rally young people together around Marxian socialism,"[1] it was many years before he fully comprehended the theoretical structure of Marxism or understood all of its political and philosophical ramifications. Unlike Confucius he could not announce on his fortieth birthday that he had reached "the age free from vacillation," but only that he was prepared for a long period of effort to master the Marxist system. Yet, Kawakami gained recognition as an authority on Marxism almost from the day he became a convert. If he knew little about Marxist theory at the time, most intellectuals in Japan knew less.

Marxist thought was of course not entirely new to the Japanese scene in the post-World War I period. The early Japanese socialists, such as Sakai Toshihiko and Yamakawa Hitoshi, pioneered in introducing socialist thought in the first decade of the twentieth century when they established the first Japan Socialist Party. Their early efforts to transplant Marxist thought failed, however, and with the exception of a few of the veteran socialists and a few of their highly censored translations, little remained of their ground-breaking labors.

Kawakami played a major role in post-World War I efforts to reintroduce Marxism to the Japanese intellectual community. His university lectures and his journal, *Research in Social Problems,* together with other independent magazines and journals such as *Warera,* Yamakawa's *Socialist Research (Shakai kenkyū),* and *Reconstruction (Kaizō),* helped disseminate orthodox Marxist thought by publishing translations and exegeses of basic German Marxist texts, so that by the end of the nineteen twenties, intellectuals throughout the country felt competent to elaborate upon the most subtle intricacies of Marxist thought.

Although Kawakami may have lacked the theoretical competence of other contemporary Japanese Marxists, he enjoyed greater prestige, because he was already an established authority on European thought. His proficiency in foreign languages gained him access to academic discourses emanating from Germany, England, and the United States. The great number of translations and commentaries on Western thought which he published throughout his career attest to his skill in handling foreign language materials. Kawakami's background in European economic thought and the sheer volume of publications bearing his name furthered his reputation as a scholar and, hence, increased his influence over the minds of the younger generation.

Kawakami's membership in the exclusive coterie of Imperial University professors lent further respectability to his credentials for intellectual leadership. He enjoyed the advantage denied to political activists of being able to clothe his increasingly more radical messages in the garb of objective scholarship. The two decades he spent on the Kyoto campus—the first halcyon, the second tumultuous—gave him the financial security and legal protection necessary to pursue his studies in economic thought and to explore relatively freely such "dangerous thought" as Marxism, forbidden to those outside the academic community. "I had greater freedom of speech than the popular socialists," he admitted. "Sakai Toshihiko, Yamakawa Hitoshi, and others [had their works] banned as soon as they picked up their pens, whereas I was nonchalantly writing without even taking a sabbatical leave."[2]

Respect for Kawakami's academic qualifications was further enhanced by his prose style. Even as a young man writing under an assumed name, he had dazzled his contemporaries with the forcefulness of his arguments and his command over language. His first love had been literature and it had served him well: he had made his name as a man of letters even before establishing himself as an authority on economics. Facility with language and pen many times proved sufficient to win readers to his point of view, especially after the Meiji period, when he substituted a more vernacular prose for the formal Japanese style of writing.

Most important, however, in explaining Kawakami's attraction for the younger generation are the ethical qualities he embodied. His depiction of Marxism as a science of distribution suggested an exciting new academic career in humanist economics for the youth of Japan, and his own dedication to solving the problem of poverty through economic research became an example for them to follow. He related their academic studies to Japan's social problems in ways that previous economists had failed to do. Inspired by Kawakami's devotion to Marxist studies, young students flocked to the university to study under his tutelage, seeking in his lectures guidelines for political practice.

Yet, the irony of Kawakami's life is that he could never fully accept the Marxist world view in its entirety, as his students and the older generation of Japanese socialists willingly did. Kawakami's road to Marxism was tortuous. His struggles with intellectual opponents, with students, with himself; his agonizing sessions with Marx's *Capital*; and his successive conversions and about-faces all document the torment of his indecision.

Kawakami's turn to Marxism was a natural outgrowth of his previous academic interests and intellectual orientation. His education in German economic thought, his longtime concern with the problem of poverty, and his condemnation of economic individualism paved the way for his receptivity to Marxist critiques of bourgeois economics. However, Kawakami retained his antipathy to certain aspects of early Japanese socialism, and this critical attitude extended to Marxism as well. To those who do not share the Marxist world view, it should not be necessary to explain why one intellectual had difficulty embracing it. Nevertheless, the intense efforts Kawakami made to scale the walls of scientific socialism and, after many years of trying, his belief that he had succeeded, surely justifies the depiction of his labors.

Doubts about Marxism stemmed first of all from fundamental differences between Kawakami and other Marxists over methods of reform. In Kawakami's view, scientific socialists wrongly neglected moral reform in their exclusive emphasis on institutional change. This remained one of Kawakami's major criticisms of Marxism.

In addition to philosophic disagreements, however, practical intellectual problems also stood in the way of Kawakami's total acceptance of Marxism. Lacking a background in economic theory, Kawakami found it difficult to evaluate Marx's analysis of capitalism. He was generally familiar with the ideas of British economic thinkers like Adam Smith, Ricardo, and Malthus, whose work served as the starting point for Marx's thinking about economics, but he had not mastered economic theory. Although at least five decades separated the publication of Adam Smith's *Wealth of Nations* from Karl Marx's earliest writings, Kawakami's studies of Marxism proceeded almost simultaneously with his education in classical economics.

Much to his credit, Kawakami was quick to confess his weakness as an economist and to accept instruction from others. His scholarly attitude is well documented in his correspondence with Kushida Tamizō. "I don't question the study of commodity value in . . . the first chapter of the first volume of Marx's *Capital*," he wrote in one letter. "Such an explanation, I presume, is similar to your thesis. Only, what I question is whether or not commodity value = value; therefore, no value outside commodity value."[3] Later in this same letter, expressing dissatisfaction with one of Kushida's explanations, Kawakami humbly wondered, "Am I being immature in my thinking by asking why?"[4] "Fortunately, I am still not inflexible," he told Kushida, in 1924. "I want to move forward on the basis of the criticism I received from you. I believe many problems have been presented, and I keep going along gradually revising my ideas."[5]

These reservations about the validity of Marxist theory prevented Kawakami from accepting the economic analysis presented in *Capital* until he thoroughly understood it. "*Capital* is either entirely correct or entirely wrong," one of his students quoted him as saying. "It is not partly correct and partly wrong."[6] Although he introduced material from *Capital* in his course on economic theory beginning in 1919, it was not until 1927 that he reorganized his lectures entirely according to the theories formulated in Marx's magnum opus.

Another factor complicating Kawakami's efforts to comprehend Marxism was his lack of familiarity with the German philosophical tradition that had nurtured Karl Marx and had contributed the crucial concept of the dialectic to Marx's view of history. Kawakami's study of Marxism involved him not only in a parallel study of British economics but also plunged him, eventually, into the mystifying abstractions of Hegelian philosophy. Concentrating at first mainly on the economic arguments in *Capital*, however, Kawakami avoided the philosophical implications of the materialist conception of history lying at the basis of the Marxist system.

This neglect of historical materialism became the central issue in Kawakami's debates with his critics in the early nineteen twenties. The debates focused on the meaning of historical materialism, but they extended to the larger issue of the role played by moral men in shaping human society. To appreciate these emotional exchanges between Kawakami and other Japanese Marxists, it is necessary first to define briefly Marx's materialist conception of history.

The Doctrine of Historical Materialism

By repudiating the Hegelian thought system, Marx claimed to have put an end to philosophy. What he meant by philosophy, however, was philosophical idealism. His own materialist view of history, which considered concrete social facts rather than consciousness as the primary reality, was nevertheless closely bound up with the German speculative tradition.

Marx applied Hegel's law of the dialectic to human society, demonstrating how the unfolding of necessary class antagonisms supplies the driving force of historical progress. The clash between economic classes—that is, between those who own the means of production and those who labor under the owners—is the motor of change from one stage of history to the next. These class antagonisms erupt into open violence, or revolution, at a certain stage in the development of productive forces. In the final stage of this historical process, the exploited working class seizes political power and, taking the means of production into its own hands, puts an end to class conflict. Communist society is presumably a classless society, in which the means of production are owned by all or at least a majority of the people.[7]

In this interpretation of the historical process, philosophical systems, law, religion, art, and even economic theory are seen as expressions of the ideologies of economic classes. Marxism supposedly ripped the mask off philosophy, revealing its true identity as the rationalization of class interest. It is in this sense that Marx claimed to have replaced philosophy with science. His science was a philosophy of history, written from the materialist point of view.

Marx was obviously using the word science (*Wissenschaft*) in a special way, to designate the study of man's real life in society, that is, the economic nexus that defines man's existence.[8] Thus, Engels credited Marx's discovery of the materialist conception of history with having transformed socialism into a science, because it placed socialism on a real basis, which is to say, on the basis of economic interests. The founders of the various schools of utopian socialism, Engles wrote, had criticized capitalism in the name of absolute truth, reason, justice, and other familiar Enlightenment ideals, but each man's version of these abstractions had been conditioned by his own subjective understanding.[9] The doctrine of historical materialism transformed utopian socialism into a science precisely because it demonstrated that ideals were no more than class ideologies and that what propelled history forward were changes in the means of production.

The classic formulation of this theory of history is found in Marx's preface to *A Contribution to the Critique of Political Economy,* one of the many Marxist writings Kawakami read between 1919 and 1920:

> In the social production of their life men enter into definite relationships that are indispensable and independent of their will, relations of production which correspond to a definite stage of development of their material productive forces. The sum total of these relations of production constitutes the economic structure of society, the real foundation, on which rises a legal and political superstructure and to which correspond definite forms of social consciousness. The mode of production of material life conditions the social, political, and intellectual life process in general. It is not the consciousness of men that determines their being, but, on the contrary, their social being that determines their consciousness.[10]

What is important for the present discussion are the philosophical implications of the doctrine that was supposed to have put an end to philosophy. According to this presumably scientific study of history and society, goals could not be set by seemingly well-intentioned individuals, nor policies formulated by government leaders or social reformers, each acting on the basis of his own particular set of values and interests. On the journey from utopian to scientific socialism, Marxists had dropped the baggage of absolute morality and universal ethics, piecemeal reform schemes and philanthropic gestures, calls to individuals and cries to the government for help. It is the spontaneous workers' movement that carries change forward, through revolution.

> At a certain stage of their development the material productive forces of society come into conflict with the existing relations of production or—what is but a legal expression for the same thing—with the property relations within which they have been at work hitherto. From forms of development of the productive forces these relations turn into their fetters. Then begins an epoch of social revolution.[11]

The doctrine of historical materialism, thus far summarized, appears to be a description of a purely amoral, determined historical process. Nevertheless, Marx's own words to the contrary are familiar even to those who have never studied his writings: the task of the philosopher was not to interpret the world but to change it.[12]

Historical materialism has its own ethical implications, predicated on the belief that history is moving toward the realization of a rational and moral society. In this sense, Marx carried a heavy, though unacknowledged, debt to Hegel. The proletariat is the first class to become conscious, or at least partially conscious, of those seemingly blind economic laws which govern historical development, and their understanding of these laws enables them to achieve rational control over them. The tension between determinism and human volition, writes George Lichtheim, is "discharged on the plane of action, through practical revolutionary manipulation of those very 'historical forces' which appeared on the theoretical level as blind instruments of an impersonal destiny."[13]

In Marx's theory of history, the moral realm of human consciousness and the material realm of economic laws are intimately related through the act of revolution. In showing how they are related, Marx laid claim to having established a science of history. His studies of capitalist institutions followed and were based upon this new way of viewing history. Without recognizing the assumptions about history that he brought to his work, it would be impossible to understand what is meant by science in Marxism. Accepting the materialist conception of history demanded more than the positivist's pledge to study objective social facts. It demanded above all accepting the description of the historical process in terms of class struggle and cultural life in terms of class ideology. The inability to make this commitment to the Marxist world view eventually singled out Kawakami as an unorthodox, "special,"

Marxist, struggling to balance the science of Marxism with his own ethics of selflessness.

Kawakami's Treatment of Historical Materialism

Most of Kawakami's articles in his new journal, *Research in Social Problems,* between 1919 and 1920, were devoted to introducing the rudiments of Marxist thought. In the first ten issues of the journal, Kawakami explained such basic Marxist concepts as class war and the labor theory of value. During this year and for several years afterward, while he was sorting out in his own mind the differences between utopian and scientific socialism and between classical and Marxist economics, Kawakami's sources were primarily Western and especially German, and his interest lay in Marxism as a general theory, rather than as an economic interpretation of Japanese history.[14]

In Kawakami's initial attempt to fit together the pieces of the Marxist puzzle, he called historical materialism one of the three essential principles of Marxism. The other two principles were the surplus theory of value and socialism. According to his explanation, the "gold thread" connecting these three principles was the theory of class war.[15] Elsewhere Kawakami defined historical materialism as a "causal law of historical progress" and a "kind of theory of inevitability," which predicted the eventual downfall of capitalist society. "Marx discovered the scientific law of the cause and effect of historical progress. The causal law is called historical materialism. Historical materialism is the basis of Marx's scientific socialism."[16]

Marx's view of history was scientific, Kawakami explained, because it showed that the realization of socialism was intimately related to the development of economic forces. Unlike the utopian socialists, Marx did not simply create an ideal society out of his head; rather, he showed why socialist organization was inevitable. Whereas the utopian socialists invented the ideal society, Marx discovered the conditions within present society for its realization. The difference between utopian and scientific socialism was that the first was a theory of will, while the second was a theory of fate.[17]

Did historical materialism suggest that human will was entirely insignificant in affecting the course of history? In a 1919 article about the "Evolution of Socialism" Kawakami agreed that men would have to entrust more to historical laws and less to individual actors. Social reform could not be accomplished by one or two men but depended upon the ripening of class consciousness, which in turn depended upon changes in the means of production. On the other hand, scientific socialism provided new methods for reaching socialism. In place of a handful of moral do-gooders, like Robert Owen, Marxism drew large numbers of workers into political movements, increasing the chances for success. Although scientific socialism replaced faith in individual will with a theory of causality, it also appeared to have strengthened practical reform movements by offering concrete political tactics aimed at organizing the proletariat for the purpose of seizing power. These measures were not necessarily violent; rather, they consisted of first gaining the right of suffrage and then, gradually and constitutionally, forming parties representing proletariat interests, electing members to the parliament and eventually seizing power through peaceful means.

The difference between utopian and scientific socialism, in short, was that the first was a religious movement engaged in preaching and educating, whereas the second was a political movement, based on the scientific reading of history.[18] Historical materialism was responsible for this difference, for it separated "what is from what ought to be." It was necessary first "to have factual understanding upon which to base your ideals. No matter how lofty, ideals not grounded in scientific knowledge are the utopias of the world of dreams and cannot reform actual human life."[19] Scientific socialism demonstrated, in Kawakami's understanding, the need for reformers both to study the economic laws of history and to participate in practical political movements.

The scientific socialists' denunciation of utopian socialism as naive and impractical echoed Kawakami's own objections, voiced some fifteen years earlier. In 1905 Kawakami, then a young economics professor, had wondered how it was possible to achieve the Japanese Christian socialists' ideal society. On the other hand, although scientific socialists had increased the prospects for realizing socialist society, they had robbed socialism of its moral nature, Kawakami complained, by replacing ethical teachings with political tactics. Repeating his longtime criticism of the non-Christian Japanese socialists for practicing a "kind of shallow materialism in which they neglect religion and show contempt for morality," Kawakami chided the scientific socialists for thinking that institutional change alone would bring about the socialist goal. Quoting Arthur Henderson, the Secretary of the British Labor party, he warned that no attempt to build a new society would succeed unless efforts were also made to raise men's moral level so that they would dedicate themselves to the public good.

> Many of today's scientific socialists have a tendency to consider the reconstruction of economic organization alone as their sole aim. They view all men as selfish ... Men are generally selfish; however, there exist some righteous men [*shishi ninjō*] who dedicate themselves to the public good. You may regard the common people who comprise society as things, animals, or mechanical beings, but all human beings possess a heart [*kokoro*] of benevolence and right-

eousness [*jingi*], no matter how small it may be. When we view human beings in this way, all men are possessed of heart and consciousness. Therefore, some men hold the view that one should engage in social reform that is based on the cultivation of men's hearts, in addition to reform based on external coercion. The earlier socialists emphasized the former and ignored the latter, whereas the contemporary socialists tend to fall into the trap of overemphasizing the latter and ignoring the former. Science does not contradict religious morality. Scientific socialism—the materialist historical view—does not contradict religious morality either.[20]

This insistence on the need for moral as well as institutional reform, a fundamental tenet of Kawakami's pre-Marxist world view, became the major stumbling block in the way of his total acceptance of historical materialism. His two-pronged attack on social problems reflected his more general conviction that economics and moral philosophy should be reunited, so that science would be placed in the service of justice and virtue. This fervent desire to reconcile ethics and economics lay at the very heart of his interest in socialist thought, coloring his interpretation of the past as well as his vision of the future.

Kawakami's own interpretation of the evolution of socialism owed little to the materialist laws of history. In his lectures on economic theory at Kyoto Imperial University throughout the nineteen twenties, Kawakami explained the rise of socialism by narrating the moral awakening of economists, rather than the working of economic forces. The play of ideas within the minds of men, and not the unfolding of contradictions between economic classes, had contributed, in his explanation, to the dawn of the socialist era. Marx was the last in a long line of European economic theorists, but the heroes in Kawakami's historical pageant were John Ruskin and John Stuart Mill, whom he treated as forerunners of socialism, because they had led the fight against classical economics by repudiating the private profit motive.[21]

The economics of egotism, in Kawakami's interpretation, began with "The Fable of the Bees," Bernard Mandeville's cynical inquiry into the origin of virtue; continued with Adam Smith's *Wealth of Nations;* and reached its height with Jeremy Bentham's utilitarianism.[22] As the spokesman and formulator of individualistic economic theory, Adam Smith had assumed that men acted out of selfishness, but that their self-interested acts, through an invisible hand, redounded to the benefit of the public good. Smith's ideas had helped free the entrepreneurial class from governmental restrictions and enabled it to prosper, but the Scottish economist had not lived long enough to see the evils of industrialism.[23]

Although nineteenth century economists after Smith's time and even before Marx came to see the fallacy in Smith's optimistic account of capitalist laws and even agreed that harmony between capital and labor under capitalism was impossible, they failed to offer any alternatives to Smith's economics. Malthus called attention to the problem of poverty, but he believed that government relief programs would only interfere with natural checks on population increase, such as famine and infanticide, which he saw as the only two ways to control population growth. Recognizing the existence of social problems, both Malthus and Ricardo nonetheless insisted that government intervention, conscious policy planning, or moral restraint on the part of the rich to rectify social ills would only make matters worse, though they conceded that moral restraint on the part of the poor to check population growth might have some beneficial effect.[24] Malthus' basic thought, Kawakami concluded, was that "distribution based on the inequality of wealth was necessary for the progress of mankind."[25] In the society based upon individualism, he added, "each member . . . is given freedom for his own economic activity, but at the price of freedom which he has all to himself, . . . he is to be responsible for his own economic fate."[26]

Men's bondage to economic forces over which they seemingly had no control was finally broken by John Stuart Mill, who made the important discovery that the distribution of wealth, unlike its production, depended on the laws and customs of society, and not on certain iron laws of nature with which men could not tamper. Mill argued that it was possible for government to intervene in order to effect a more equal distribution of wealth, because the question of how income was distributed depended ultimately on how men believed it ought to be distributed. If one thought that economic inequities should be rectified, then one could design methods to equalize wealth.[27]

Socialism was an idea, a moral breakthrough that had permitted a subsequent scientific breakthrough in economics. Faith in man's reason and ultimate goodness confirmed Kawakami's belief that men were free to create the society they deemed desirable: men could remake their own world, on the basis of the goals they had chosen. This belief explains the significance in Kawakami's writings of Mill, Ruskin, and Marx, "humanist economists" who had shown that capitalism could be changed through the conscious repudiation of "individualist economics." The reasons why economic thinkers like John Stuart Mill had finally arrived at such a new moral consciousness fascinated Kawakami. In his lectures he was fond of relating how Mill had broken away from his father's stern discipline and utilitarian economics and, under the influence of the warm personality of Harriet Taylor, had drifted close to socialism.[28]

Kawakami resisted the Marxist notion that men's thought and actions were determined purely by economic considerations. He did not attribute a higher historical morality to any particular class, nor did he treat the emergence of socialism as an outcome of contradictions within the economic order. Rather, he continued to locate the cause of social change in the ethical awakening of individuals to the need for institutional reform.

An eagerness to view socialism as the product of human conscience, rather than the outcome of class warfare, was understandable in the light of Kawakami's personal preoccupation with moral values and spiritual self-development. It was precisely his lingering idealism, however, which earned for him the scorn of the more thorough-going Japanese Marxists. Kawakami's reference to himself as a scientific socialist who affirmed the existence of an unchanging morality revealed that he had not yet accepted the economic determinism and moral relativity that lay at the base of Marxist thought.[29] His definition of that unchanging morality further antagonized his critics, not only because he freely confessed the religious roots of his "ethics of absolute selflessness," but also because the application of this ethics to individual behavior would have condemned the Japanese to the kind of self-negating, feudal values on which the Japanese state had traditionally relied for its authoritarian control.

Kawakami's efforts to play the sage and scientist at once earned the scorn of veteran Japanese Marxists like Sakai Toshihiko, who criticized him for his "incomplete blend of humanist philosophy and socialist economics," complaining that Kawakami suffered from the "difficult-to-renounce disease of humanism" (*jindōshugi*).[30] Sakai's criticism sparked a vigorous debate touching on matters of crucial importance for the early Japanese understanding of Marxism.

Humanism and Science in Marx

The issues dividing Kawakami from his critics in the early years of Japanese Marxism centered on the meaning of humanism in Marxist thought and its relation to the presumably scientific causal laws of history. If Marxism in Japan introduced a new ethical consciousness enjoining intellectuals to work for the welfare of the poor, it also introduced a new economics—the science of distribution—and a new, supposedly scientific way of viewing history as well. What was the relation between the humanist and the scientific elements in Marxism?

Socialism was infused with a humanist philosophy. Surely Marx's tone of moral indignation and the almost messianic promise of deliverence in some of his writings were elements as strong as, if not stronger than, his scientific treatment of history. How did these impassioned pleas for men to free themselves from oppression fit into historical materialism?

This question posed a problem of logical consistency. If Marxism was a science, it had nothing to do with ethics and, therefore, the Marxist theoretician was totally justified, indeed, required by logic to distinguish between a normative philosophy and the causal laws of history. If it were true, as one critic of Marxism had said, that "in all of Marxism from beginning to end, there is not a grain of ethics," how then could socialism be defended in moral terms?[31]

The debates essentially boiled down to the question of whether there was indeed an ethical aspect to the science of Marxism and, if so, whether this aspect was similar to Kawakami's humanism. Obviously the word humanism was used in the general sense of idealistic, that is, anything concerned with values, morality, religion, the spiritual realm, and ultimately philosophy, whereas science implied economic forces.[32]

Several months after Kawakami began publishing *Research in Social Problems,* Sakai Toshihiko forcefully refuted his view of Marxism as an altruistic ethics. In an article entitled "The Most Fearful Defects of Modern Socialism," published in the June 1919 issue of the new magazine *Liberation,* Sakai contended that Kawakami's repeated references to man's moral perfection implied the belief in an unchanging morality and typified pronouncements of the power class. Sakai, the seasoned materialist, reviewed the essentials of Marxism: all thought was merely ideology, changing in conjunction with changes in the material means of production. An unchanging morality simply did not exist.[33]

In defense of his own position, the following month Kawakami published an article on "Changing Morality and Unchanging Morality" in which he attempted to explain that, although the content of morality changed with different historical eras, an underlying continuity in moral thought existed.

> I recognize the evolution of morality. In certain times and, moreover, in certain eras, cannibalism was considered morally permissible. Slavery, too, was considered morally permissible. However, today these things are morally reprehensible. Similarly, today's society considers it morally acceptable to hire human beings as laborers. According to materials recorded in Dr. Shimamura Ikuto's *Studies in Relief,* the number of customers of one prostitute in one month reached ninety-two men . . . We treat human beings like slaves in this way, and in Taishō Japan, it is even legal . . . However, I believe that we must inevitably reach a time when this morality and these laws too will change. Just as cannibalism and slave systems today are repudiated, so today's wage system and licensed prostitution, in the future society, will necessarily be rejected. In this sense, I recognize the evolution of morality . . . At the same time, however, I believe that morality is unchanging . . .
>
> Mr. Sakai says my understanding of the origins of morality and the

evolution of its content is very inadequate . ., Of course there are contradictions and inconsistencies in my thought. But, if my "sickness" is to believe in an unchanging, eternal, and absolute truth, then I can never part completely from this sickness.[34]

In an effort to mediate between Sakai and Kawakami, Kushida Tamizō suggested that Marxism rested upon two separate bases: the philosophy of historical materialism and the ideal of the liberation of mankind (revolution). These two bases did not necessarily contradict each other: the first was seen in the realistic writings of Marxist economists and in Marx's *Critique of Political Economy;* and the second, in the idealistic writings of the Communists and Marx's *Communist Manifesto*. If Kawakami's unchanging morality referred to Marx's ideal of the liberation of mankind, then he was justified in using the term, because in the sense of liberating all those who are immiserated by the social conditions of the day, Marx's goal might be construed as an absolute ethics.[35]

After further research and reflection, however, Kushida published in 1920 a review of Kawakami's latest book, *A History of Modern Economic Thought (Kinsei keizai shisō shi ron)*,[36] in which he indicated that he was no longer willing to consider Kawakami's ethical fervor equivalent to Marx's revolutionary fervor. By this time Kushida had resigned his position as an instructor in economics at Tokyo Imperial University to join the staff of the Ohara Research Institute, a private, social science research organization, and had devoted himself full time to Marxist studies. He now denied that humanism was one of the two pillars supporting Marxism and suggested instead that historical materialism was the sole basis upon which the structure of Marxist thought rested. There was no absolute ethics in Marxism, only the morality of a specific economic class, carrying out its historic mission. The liberation of the working class would come about not out of the triumph of moral outrage or an ethics of absolute unselfishness but through the operation of the inexorable laws of history. "The method of reaching each new landmark," Kushida asserted, "is none other than the workers' movement," and the only "scientifically possible policy" was class warfare.[37]

It had become clear to his critics that Kawakami's humanist vocabulary owed more to Confucius than Marx. Where Marx spoke of freeing men from class masters, Kawakami spoke of liberating them from their selfish desires. Where Marx wrote of self-fulfillment, Kawakami wrote of self-sacrifice. The young Marx dreamed of a society of "whole men," as Richard Tucker writes, "well-rounded men," finding fulfillment in self-gratifying, creative work, freed from economic bondage to enjoy the full "totality of human life-activities."[38] In contrast, Kawakami summed up the goal of human life in the Confucian words "hearing the Way." Human life was a journey toward that state of moral perfection in which "one wants to do what one ought to do." Ultimately, the differences between Kawakami and Marx reflect the philosophical traditions which nurtured them. Marx's spirit of individualism betrays the influence of the French Enlightenment; Kawakami's altruism demonstrates the powerful grip on him of Japanese morality.

Reluctant to criticize his former teacher and close friend—the man who had helped launch his career—Kushida had tactfully begun his review by telling his readers he would merely try to explore areas of doubt within his own mind. But although he had written the article ostensibly to shadowbox with his own doubts, he ended by challenging Kawakami and winning a clear victory. One month after Kushida's essay appeared, Kawakami printed his public confession of defeat, dedicated to Kushida and bearing the somewhat melancholy title "On Man's Self-Deceiving Nature." He admitted he was guilty of utopian dreaming and vowed to become a realist, by studying historical materialism, whose "basic spirit is realistic."[39]

Kushida had convinced Kawakami that an understanding of Marxist philosophy—the dialectic applied to material society—was the prerequisite for his studies in Marxist economics. In the following two years, while Kushida was abroad studying in Germany, his teacher-turned-pupil undertook the study of philosophy, going back to Kant and Hegel to get a running start on Marx.

Kawakami found the study of European philosophy extremely tedious. Unschooled in the European philosophical thought his own students were receiving, he felt helpless before the tomes on German idealism that filled his study. "I was annoyed by the young people saying, 'Kant, Kant,' so this vacation I planned to read Kant," he confided in a letter to Kushida. "Since yesterday I've been looking at *The Critique of Reason* but, you know, it's extraordinarily difficult ... Eating is necessary in order to live, and so we have all come to eat with ease, but I wonder if either Kant's philosophy or Hegel's philosophy is necessary." Although he believed that his humanist thought provided an adequate philosophical basis for Marxism, he realized that, without a persuasive modern definition of his philosophical position, he could not continue for much longer as an authority of Marxism. Kawakami often quoted Christ's words about man not living by bread alone, but now somewhat wryly he added that "if philosophy is necessary in order to live, we will die before understanding philosophy."[40] For a man of frail health and sensitive disposition, there was perhaps more truth than jest in the comment.

8. historical materialism and revolutionary will

In 1920, Kawakami Hajime became chairman of the faculty of economics at Kyoto Imperial University. A year earlier, economics had been established as a separate college independent of the faculty of law, a change reflecting the growing recognition of economics as a specialized branch of learning. Largely responsible for popularizing economic studies and widely respected by his colleagues and students, Kawakami Hajime seemed an appropriate choice to head the new faculty, though he himself viewed the appointment with mixed feelings, because much of his time was now consumed by administrative duties which, together with his teaching responsibilities and the publication of his monthly journal, *Research in Social Problems,* kept him working until late into the night. Although the journal ran to only about thirty pages, it took him two days to write one issue.[1] All this work, but especially his administrative duties, took a great toll on his health and mental well-being. "I have spent all my time from morning to night every day on the matter of the independence of the economics department," he complained, in a letter to Kushida Tamizō.

> In the evening I've been so exhausted, I haven't been able to pick up my pen, so I have neglected to write to you. It is very tiring to do this kind of work because it is so unfamiliar to me. The effort and agony I put into academic matters is quite different from what goes into business matters. Over two weeks have gone by without my picking up a book; this kind of thing is very rare in my life, but I do it, consoling myself with the thought that the sacrifice is unavoidable at this point.[2]

One of the few encouragements in these otherwise disturbing years was the unexpected success of his journal. When he first launched *Research in Social Problems,* on New Year's Day 1919, he had considered it "a tiny little magazine as small as a grain of pepper thrown into a desert." Nevertheless, within six months circulation had reached over twenty thousand copies. Gratified, Kawakami wrote on the back cover of the third issue, "I did not expect to find such a small work like this would be picked up by so many sympathetic people and so widely disseminated."[3]

Not all of his readers were sympathetic, however. "Before long," he recalled, "the journal became a hot issue for the ruling class,"[4] and he confronted the heavy hand of the censor. Words such as class war or revolution became particularly susceptible to excision as the decade wore on.

It was mainly on the pages of his journal that Kawakami published his latest translations and interpretations of Marxism. An examination of these journal articles reveals that, between 1920 and 1924, Kawakami was still primarily engaged in explaining the component parts of European Marxist thought. Nevertheless, his frank discussions of revolution and socialist society represented, at least for him, a major departure from his previous concentration on moral reform and spiritual reconstruction. By 1920 he had halted publication of *Tale of Poverty* and, in the revised edition of his *Views on Current Problems,* originally published in September 1918 and reissued in 1920, he omitted articles on Ruskin and Smart and added instead articles on the "Mission of the Workers' Movement" and the "Establishment of Surplus Value."

During these years after his pledge to study German philosophy, Kawakami continued to wrestle with the meaning of historical materialism and the problems it posed for him as an ethical economist. Although he seemed willing to accept, at least in theory, the economic basis of historical change and the need for class struggle, he continued to believe in the vital importance of man's ethical consciousness in remaking society. He was particularly interested in Marx's explanation of the role played by human will in social transformation and in the crucial question of why and when men decide to overthrow the existing social order. If socialism was not the result of ethical awareness, what then created in men the will to overcome the capitalist economic order? And what enabled men to work together harmoniously after the socialist revolution? In Kawakami's view, social reconstruction alone was inadequate to gain the willing compliance of society's members, just as law and policemen alone did not assure communal cooperation. Religion and morality were necessary to help maintain a spirit of mutual good will. Therefore, spiritual reconstruction remained the "prerequisite for compulsory reconstruction." Spiritual preparation preceded structural reformation.[5]

Despite this insistence on the need for a moral revolution to accompany the political revolution, Kawakami tried to cast his position within a Marxist framework. In two books, *Social Organization and Social Revolution (Shakai soshiki to shakai kakumei)* and *Research in Historical Materialism (Yuibutsu shikan kenkyū),* Kawakami drew on the writings of Marx and the German Social Democrats, as well as Russian Marxists like Trotsky, Lenin, and Plekhanov, to ponder one of the most fundamental philosophical questions involved in Marxist thought—the relation of human consciousness to the materialist laws of history.

Voluntarism and Determinism

Kawakami's writing in this period reflects his troubled awareness of the way Marxism in the hands of Engels and Kautsky had been transformed from a critical theory to a science of causal laws. The two men had worked hard to invest Marx's philosophical materialism with the authority of a natural science. The further the orthodox Marxists moved in the direction of a natural science, however, the less chance they had to justify the socialist revolution in moral terms or to participate in its creation. As soon as historical materialism was treated as a science of the causal laws of history, the tension between determinism and human will was lost.[6]

This transformation of Marxism into historical science was reflected in Kushida Tamizō's studies after 1920. Kushida's interests were largely academic. He spent the last ten years of his life, before his death in 1934, engaged in studies of Japanese agriculture. Kawakami had always preferred stories of human self-sacrifice to tables on economic statistics, and he was energetic in dispelling the uninspired fatalism which late nineteenth century interpreters had virtually imposed on Marxism.

Determinist philosophies offended Kawakami for the obvious reason that they offered no independent role to the human actor and hence no possibility for ethical behavior. It was not that he was impatient for revolution but rather that he was too much a chronicler of men's moral life to accept a purely abstract version of the natural evolution of history. However much he struggled to achieve scholarly objectivity, his scientific studies were predicated upon the belief that these studies would inspire men to reform society. Self-sacrifice, duty, restraint—these essential virtues in his code of conduct—suggested his faith above all in the power of human rationality to correct social injustice. Persuaded that ethical consciousness had no place in the Marxist system, Kawakami turned to revolutionary consciousness as the locus of morality in Marxism. Recognizing that Marxism combined the practice of revolution with the study of economic laws, however, he attempted to explain the intimate connection between the two.

Kawakami found two seemingly contradictory, "alternating currents" in Marxism: the first, evolutionist, appeared in the *Critique of Economics;* and the second, revolutionist, inspired the writing of the *Communist Manifesto.* German critics of Marx, such as Sombart and Tönnies, had questioned how these two views could be reconciled. In his *Critique of Economics,* they argued, Marx had said that changes in social organization occurred only when the means of production reached the highest possible level of development. Yet, in the *Manifesto,* Marx had sought to arouse the will of the proletariat to seize power, even though economic conditions were not ripe for change.[7]

Kawakami offered two answers to this charge that Marx was inconsistent. The first was based on his reading of Engels' memoirs. When Marx wrote the *Manifesto,* according to Engels, he believed that the clash between productive powers and productive relations had developed to a point where revolution was possible. His analysis of conditions was wrong, Kawakami explained, but his theory was nonetheless correct.[8] "No matter what faults accompany today's capitalist organization, to the extent that there is still room for the development of society's productive power under that organization, there is proof that it still serves a function for society. Thus, a plan to overthrow capitalism would naturally end in failure."[9]

At the same time, the transition from one stage of history to another ultimately required political revolution. Even though some of the changes Marx expected to take place after the proletariat revolution, such as nationalization of the railroads, could also occur under capitalist regimes, Marxists, as distinct from other kinds of socialists or social reformers, insisted on the necessity of political revolution. Unless the proletariat seized political power, the ruling class would resist reform measures designed to transfer ownership of the means of production into the hands of the masses.[10]

The act of revolution and the socialist measures enacted after the revolution demonstrated to Kawakami's satisfaction that Marxism was more than a theory of inevitability describing natural phenomena. To be sure, much of the historical process delineated in the materialist conception of history was not subject to human manipulation, but this was true of the world as described by individualist classical economics as well. At least in Marxism, at a certain point, it became possible and indeed critically important for men to intervene actively on their own behalf. The efforts of the working class were required to destroy the old society and to build the new one:

> That society where the development of productive power is oppressive will inevitably collapse, but since ... social organization consists of the association of individuals, changes in social organization are different from natural phenomena, and *both the construction and the destruction must be done by human power and deed* ... It is not a natural change.[11]

Was violence necessary in order to seize power? Marx and Engels were ambivalent on this point. In his later years, Engels seemed to think that the proletariat could gain power legally, through political organization,[12] and Kawakami agreed, though he also believed that, when all other peaceful methods failed, violence was justified.[13] In Marx's own words, Kamakami wrote, "Whether or not blood is spilled, at some time the day must come when the proletariat seizes political power."[14]

Political revolution demonstrated the crucial role played by conscious human actors in the drama of history. Kawakami was careful to underline Marx's conviction that man makes history and that, despite the relentless operation of irrevocable laws, in the end it is up to human actors, whether oppressed workers or bourgeois leaders, to bring the socialist state into existence through some kind of deliberate effort.

In this balance between activism and determinism which Kawakami attempted to maintain, he was perhaps truer to the spirit of the young Marx than were the so-called orthodox Marxists who, with their natural laws of history and their models of mechanical inevitability, had converted Marxist theory into a lifeless classroom science. "*History* does *nothing*," Marx raged, in *The Holy Family*,

> It "possesses *no* immense wealth," it "wages *no* battles." It is *man*, real living man, that does all that, that possesses and fights; "history" is not a person apart, using man as means for its own particular aims; history is *nothing but* the activity of man pursuing his aims.[15]

At the same time, even the chiliastic Marx of the *Manifesto* considered the consciousness of the working class to be determined by its economic circumstances. Even when the workers consciously and actively participate in revolutionary practice, they are governed by the forces of history. History may rely upon men to carry out its plan, but these men have been molded, in turn, by history. Their consciousness is an awakened class consciousness. Their mental life is determined by their economic class, and their conscious activity consists in choosing to do what must be done. Even though the outcome is assured, the workers "fight for their freedom," in Isaiah Berlin's words, "not because they choose, but because they must, or rather they choose, because they must: to fight is the condition of their survival."[16]

The meaning and role of consciousness is perhaps the most controversial issue in all Marxist thought. Consciousness, the factor of human will, has served many as an entering wedge into the strictly determinist interpretations of Marx's philosophy of history. It has opened the door to idealist renditions of Marxism, in which legal, constitutional methods of achieving reforms obviated the need for the violent and abrupt overturning of the capitalist system. By the time Kawakami came to his studies of Marxism, revisionist philosophies on the European continent had provided an alternative to fatalism, and even Marx had once agreed that socialism would come by the ballot in the United States.[17] The increasingly wider areas opening up for the exercise of human choice—a choice defined not merely as consciousness of necessity—conceivably encouraged Kawakami to believe that the human spirit was as important as economic forces and even that the former was not necessarily determined by the latter.

If one were to summarize his many articles on this subject written in the first four or five years of his education in Marxism, Kawakami's somewhat eclectic version of the balance between determinism and human will would be something like the following: the breakdown in the capitalist system creates in men an ethical awareness of the system's inequities and stimulates their innate spirit of resistance, which Kawakami termed the will to live. Of course, men do not undertake what they consider to be hopeless tasks. They would not awaken to the injustices and oppressiveness of the economic system unless that system were on the verge of collapse, at which time its economic and moral inadequacies would become manifestly clear to the oppressed, who would then lay plans for change.[18]

Impetus for change was not necessarily limited to the proletarian class, however. As Kawakami began to take an interest in the Bolshevik Revolution, his hall of heroes expanded to include the category of noble revolutionary leader. He wrote more as a eulogist than as an economist when he described the leader of the Social Revolutionary Party in Russia at the time of the 1905 uprising: Gregory Gershuni "trusted the people; he loved the people. He wanted to concentrate all his energies on behalf of the people. In talking to the common man he would say, 'I feel the Russian people will doubtless accomplish the great tasks of truth and justice.' "[19]

Those men Kawakami singled out for praise were the self-sacrificing men of bourgeois as well as proletarian class origins who were in his eyes responsible for the transformation of capitalist society into socialism and whose motives were not simply determined by economic factors. To be sure, Marx also praises socialist transformers, but they are the proletariat, working on their own behalf. Their action is inspired by their awakened political consciousness, quite different from Kawakami's awakened ethical conscience, or a Social Darwinist will to live. It may be possible to avoid the mechanical fatalism of the doctrine of historical materialism by demonstrating the need for conscious actors, but one cannot avoid the fact that in Marxism, the morality of these actors is conditioned by the economic class to which they belong, and their freedom consists solely in deciding to help along their destiny.

The morality of the working class is contained within the process of history. It is an objective morality: what matters is not the subjective motives of the workers, which are admittedly selfish, but the historical mission they accomplish. At any rate, since morality is considered to be a function of economic relationships, self-cultivation of the sort Kawakami concerned himself with, would hardly be necessary in the rational socialist society. Selfishness is not an attribute of human nature. Selfishness, if one can use the term at all, is part of human history, not human nature. It is built into the very structure of historical development. "If socialist morality was essentially the expression of proletarian class interest," George Lichtheim writes, "it was evidently impossible to subordinate its aims to an allegedly suprahistorical ethic binding upon the whole of mankind. Such an ethic could be expected to evolve—if at all—only after the existing society had been left behind."[20]

Kawakami only partially submitted to this materialist doctrine. By demonstrating the importance of consciousness within the Marxist system, he tried to reserve a place in Marxism for the beneficence of human spirit. In the light of this understanding of historical materialism, it becomes possible to appreciate the enthusiasm with which he greeted news of the socialist experiment in Soviet Russia, because the Russian experience, by its obvious departure from the orthodox Marxist determinist scheme of history, curiously validated his own interpretation of the significant role reserved for moral heroes in the task of creating a socialist society.

Historical Materialism and the Russian Revolution

Although the stormy revolutionary history of twentieth century Russia, and especially the Bolshevik victory of 1917, lent encouragement to reformist efforts in Japan in the post-World War I years, interest in Russian Marxist thought among the Japanese literate public three or four years later was still slight. Translations of some of Lenin's writings had appeared in Sakai Toshihiko's journal, *New Society,* beginning in 1917; and after Sakai, Yamakawa Hitoshi and Arahata Kanson announced their conversion to Bolshevism in 1919, they began more energetic efforts to transmit Leninist thought to the Japanese. Yamakawa's journal, *Socialist Research,* contained an article in the June 1920 issue on the organization of the soviets and the proletariat dictatorship, and with the establishment of *Vanguard* (*Zen'ei*) as the new organ of the Japanese Bolshevik group, another outlet for Russian Marxism became available to Japanese intellectuals.[21]

Until Fukumoto Kazuo returned from Europe in 1924 to overwhelm the younger generation of students with his grasp of Leninist thought, however, Germany remained the seat of Marxist learning in Japan. Even when Kushida Tamizō went abroad in 1920, he chose to study in Berlin. Although he was exposed to Russian Marxism and was even invited by the famous Russian Marxist scholar Ryazanov to visit Leningrad, the books he brought back with him were mainly on German orthodox Marxism.[22] Kawakami, too, identified Marxism at first with German scholarship. In addition to Marx's original works, his reading was confined largely to the writings of Marx's German interpreters, Kautsky and Engels.

This neglect of Russian Marxism later proved a source of great embarrassment to Kawakami. He confessed that he was "the weakest of men in qualifications for leadership. Even in the reception of the spiritual influence of the Russian Revolution, I was rather late, compared to sensitive men. Despite the fact that the world's attention was drawn by the success of the Russia Revolution to a deep understanding of Marxism, that is, its philosophical basis, . . . I was still completely in the dark about it."[23]

Here Kawakami was unfair to himself, because he was actually informed of the Russian Communist experience and wrote energetically about it between 1922 and 1923. Moreover, he provided a realistic assessment of Soviet Russia's economic policies and of the implications for Marxist theory of Russian efforts to industrialize under socialism.

Kawakami's research on the 1917 Russian Revolution began in 1921,[24] about two years after his vow to study historical materialism, and in his writings around this time, as their titles suggest, he tried to fit the Russian Revolution into the framework of historical materialism: "A Dialogue on Historical Materialism—and the Russian Revolution," "The Russian Revolution and Socialist Revolution," and "The Necessity of Socialist Revolution and Historical Materialism."[25] A careful study of the theoretical implications of the Russian Revolution occupied him throughout 1922 and 1923, at the end of which time he came away more confident than ever, on the basis of the Revolution, that his special combination of idealism and materialism was a tenable interpretation of Marxism. He saw in the Soviet Russian revolutionary experience a contemporary example of Marxist laws of history which demonstrated, by its very unorthodoxy, the importance of conscious human will in shaping human history.

The Soviet leaders insisted that their revolution followed the orthodox tradition of Marxism, and in the years after 1917 Lenin and Trotsky paid homage to the ideological leadership of the German Marxist camp. Yet, for Kawakami, the most important lesson of the Russian example was that ultimately what determined whether a country moved into the socialist stage was not its level of economic development at all. "It's not—as the misinterpreters of historical materialism say—that Marxists believe that the realization of socialism occurs by a mechanical necessity," Kawakami wrote. "On the basis of the 'Russian experiment' it is possible to conceive of socialism as something attained by conscious effort."[26]

How then should one regard the theory of historical stages? Both Lenin and Trotsky had admitted that Russian society was not yet in the socialist stage, even though the proletariat had seized power. They had tried to explain this seeming deviation from orthodoxy in two ways.

First, Trotsky had introduced his concept of subjective elements to explain why the political consciousness of the Russian proletariat so far exceeded its economic predicament. In more than one essay Kawakami had occasion to quote Trotsky's words: "To think that there is an automatic dependence between the seizure of power by the proletariat and a nation's technological and industrial resources is to understand historical materialism as a very juvenile law." Such variables as a nation's cultural traditions or psychological factors could influence the degree of political consciousness among the proletariat.[27]

Second, Lenin and Trotsky quoted an obscure passage from a letter Marx had written to Wilhelm Bracke, in which he had suggested that there was a transition period between the capitalist and socialist stages. Lenin and Trotsky had seized this ideological foothold, and Kawakami, accepting their explanation, agreed with them that "today's Russia clearly belongs to Marx's transition period from capitalism to socialism."[28]

The Russian Revolution was a proper Marxian socialist revolution, Kawakami concluded, because its *goal* was socialism. Could this goal be reached, starting from the pre-capitalist relations which largely existed in Russia? Lenin himself had posed the question in his essay on the "Meaning of Agriculture Theory," the first Russian Communist work to appear in Kawakami's *Research in Social Problems*.

The transition could be achieved, Lenin had said here and elsewhere in the early days of Bolshevik rule, "in only one condition I know of—the electrification of the entire country."[29] Naive to the point of being comical, Lenin's reliance on this one scheme (summed up in his slogan "Communism is Soviet power plus the electrification of the whole country"), came to have enormous meaning for Kawakami. It confirmed his belief that socialism would not be achieved easily, or even inevitably, but required planned and purposive effort.

Whether or not the Russian revolutionaries would succeed in introducing socialism was still uncertain. Revolution when material conditions were not ripe for change might end merely in the reinstitution of old forms of political despotism. The Russian experiment as long as it did continue, however, provided Kawakami with the assurance that men could shape their own destiny. Both in the act of making a revolution and, more important, in the rational effort behind planning the new socialist society, the role of individual men, not entirely dependent upon economic forces, was paramount. Trotsky and Lenin were so busy making history, he marveled, they scarcely had time to write it.[30]

In fact, the Russian Revolution proved, if nothing else, that socialism might even be created by government leaders through certain basic economic policies. Kawakami eagerly awaited further news about the new Soviet economic programs, curious to learn how the Soviet leaders were arranging the redistribution of national income. "A revolution in economic organization," he predicted, "will end in a revolution in economics itself." Soviet Russia was taking the lead in the transition from "laissez-faire economics to state planning," and from the "study of causal laws" to the "science of goals."[31] In Kawakami's interpretation, the Soviet experiment signified the emergence of a new form of social policy, dedicated to the more just distribution of wealth.

A New Journey to Marxism

Given this understanding of events in Russia, it is perhaps easier to understand the confidence with which Kawakami published his new book in 1923 on the *Historical Development of Capitalist Economics* (*Shihonshugi keizaigaku no shiteki hatten*). This massive work, the product of his many years as an interpreter of capitalist economics, represented that idealistic approach to the study of economic history characteristic of his thinking in 1919 at the time of his conversion to Marxism. What continued to hold his attention was the evolution of socialism—the new economics of distribution—whose first application could be observed at that very moment in Russia.

The heroes of the book turned out to be, once again, the familiar personages of John Ruskin, Thomas Carlyle, and John Stuart Mill. Harbingers of the new economic age, these men had led the fight against classical economics, denying the self-seeking ethos on which the system was based. The repudiation of the private profit motive, in Kawakami's explanation, had set the stage for a new economics, dedicated to the goal of a more equitable distribution of wealth.[32]

Had Kawakami not already pronounced himself a Marxist, a realist, and a student of historical materialism, this foray into the realm of what was more intellectual than Marxist economic history might have inspired less anger in his most constant friend and critic. But the fact that Kawakami claimed to have written an economic history based on the materialist conception of history prompted Kushida Tamizō to contend that Kawakami had written instead an explanation which itself fell into the category of ideology.[33]

Kawakami never forgot how Kushida, "pounding vigorously on his big desk,"[34] delivered a furious attack on his understanding of Marxism, singling out the keystone of Kawakami's ambivalence—his faith in the "moralistic denial of private interests." Kawakami's appeal to men's generosity and sense of justice, Kushida argued, completely ignored the economic basis of those sentiments. Socialist economics itself had evolved out of the development of economic forces and represented the interests of the proletarian class. Men's existence determined their consciousness.[35]

It was all too apparent to his critic that Kawakami had not yet assimilated the various elements that comprised Marxist thought. In his five years as a student of Marxism, he had fluctuated between a materialist history of sorts and an outright intellectual or even moral history. His eagerness to view history as the creation of human conscience was understandable in the light of his personal preoccupation with moral values. But how could he endure for so many years the tension between two contradictory attitudes toward history?

It is true that Marxist theory was itself not without inconsistencies and ambiguities, and by the time the Russian revolutionaries had established themselves in power, they had twisted Marxism into something quite different from its original meaning. This explanation still does not fully account for Kawakami's confident publication of a history of classical economic thought which featured bourgeois heroes in leading roles and pushed to the background the material circumstances of their existence.

The answer lies in Kawakami's belief in the existence of two truths, a scientific truth and a moral truth, and his need to reconcile the two. He did not recognize the total intellectual commitment demanded of him as an adherent to Marxism. Unwilling to choose between an idealist and a materialist view of history, he accepted both, straddling two mutually exclusive philosophies.

If Kawakami had approached Marxism with a background in German philosophy, he might have more readily seen the significance of Marx's Hegelian revolt. He might then have appreciated sooner the crucial message of historical materialism: all thought, including moral philosophy, religion, and even economics, is no more than class ideology, the expression of economic class interests. The choices before him would not have been easier to make, only more difficult to avoid.

But Kawakami had never approached philosophy in terms of logical consistency. His role as a modern Japanese intellectual was, quite the contrary, to reconcile, to harmonize, to blend ethics and economics—in other words, his role was precisely to straddle two views of the world. He sought to create a society based on altruistic principles and guided by scientific knowledge. Mill and Marx represented the two sides of his personality and symbolized the dualism that haunted his life: humanism and science. In his loyalty to both men, he tried to close the gap between his philosophic world view and the tenets of historical materialism.

Later in his life, Kawakami came to appreciate the irony of his defeat at Kushida's hands. "Although I was at the time already considered an authority on Marxism, I began to feel, because of Kushida's criticism, that I was outside the gates of Marxist studies."[36] Accepting his criticism, indeed, "reeling back from its force," Kawakami told Kushida, "I'm beaten. You've got the better of me. I must correct my scholarship again."[37] Six years had already elapsed since he had "greeted the spring of my fortieth year in indecision," and when his book came out, Kawakami was at a summer resort, recuperating from a recent illness. Nevertheless, spurred on rather than crushed by Kushida's latest and most devastating attack, he rallied once again and firmly resolved to continue his efforts to reach the true understanding of Marxism.[38]

Out of his sense of defeat grew a new resolve to master Marxism, and he characteristically expressed his renewed determination in the form of a poem: "Without shaking off/ The dust of the last journey/ I must set out again on a new road."[39]

part 3. communist revolutionary

the professor as political activist

Paper wars with Kushida over Marxist theory had done little to diminish Kawakami's popularity among his students or to damage his reputation as an authority on Marxism. What attracted many students to his side in the early twenties, in fact, was probably the very element in his thought most vulnerable to attack from more thorough-going Marxist ideologues—Kawakami's humanism. Whereas Kushida's logic was precise and his reasoning exact, he limited himself to his work at the Ohara Research Institute, going about his studies in the manner of a scientist: dispassionate and reserved. At no time had he ever considered it his duty to "speak to the masses."

By contrast, Kawakami's prose excited both heart and head. "I was ... searching painfully for truth," wrote one Tokyo Imperial University student, "and for something which I might wholeheartedly accept as my way of life—a common phenomenon among young people. It was just after entering preparatory school that I first experienced this longing. From the spring of that year I shut myself up in the library searching in vain until the day when I came upon Dr. Hajime Kawakami's book, *The Story of the Poor*. I was touched by its unselfish love and moved by its humanism." The young man followed the monthly editions of Kawakami's journal and recalled that he was drawn to the study of communism and materialism after reading Kawakami's stories of the self-sacrificing Russian Bolsheviks. He was "deeply moved by the biographies of the heroes of the Russian Revolution ... and by the old Russian revolutionists ... who selflessly worked for their people and country." This same student, Tanaka Seigen, later led bands of militant Communists in street fights with the police.[1]

Kawakami's classes on economics, given in the largest lecture hall in the economics department, were always filled to capacity. Students from other departments and even those unaffiliated with the university attended and willingly stood in the aisles taking notes, as the frail scholar slowly read his lectures. He was exceptionally conscientious about his teaching responsibilities, a trait that endeared him to his students. In his published works and in his lectures, Kawakami consciously endeavored to simplify his explanations, in order to make his meaning clear to the young student or layman. He went over sentences, reading them any number of times, weeding out archaic Chinese characters and substituting wherever possible the Japanese writing of the word, and then rereading the finished piece out loud to make sure the sentences flowed smoothly. The result was a conversational style that was at once modest and persuasive.[2]

Kawakami's simple explanations of the complicated components of Marxist thought acted like a magnet to draw disciples to his side. Before he knew it, the Kyoto campus became a breeding ground for political activism. Inspired by Kawakami's dedication to Marxist studies, young students came to the university to study under his tutelage. A decision to study with Kawakami was tantamount to becoming a political activist, at least in the mind of the particular student making the decision. "I must take practical action," a Tokyo Imperial University student exclaims, in a novel about student life in the nineteen twenties. "After graduating, I'm going to Kyoto to receive instruction from Dr. Kawakami."[3]

Yet, there was something incongruous about a zealous young radical passionately announcing his decision to become committed or engage in practical activity, and then going off to attend the lectures of Dr. Kawakami. The radical's hero was a frail, soft-spoken scholar, so wrapped up in his research that he rarely left the confines of the campus. While students were fighting the police in the streets of Kyoto at the end of the nineteen twenties, he was still thrashing out the meaning of Marxism in the quiet of his study.

In his first few years as a student of Marxism, Kawakami's political activities beyond the gates of the university were limited mainly to speech making on behalf of the labor movement. In February 1919, at a meeting sponsored by the Kansai Federation of Friendly Societies, he addressed fifteen hundred people in Kobe on "The Need for the Recognition of Labor Unions," and in 1921 he and several members of the Kyoto Imperial University faculty publicly announced their support of the Friendly Society.[4] Meanwhile, on campus, Kawakami continued to serve as the focal point for several extracurricular reading clubs and study societies organized by students for the purpose of reading Marxist texts.[5] One of these groups met at the university once a week after dinner to read and discuss Marxist thought.[6] Since there were few adequate translations of German Marxist writings at the time, Kawakami helped the students read works in the original German, translating and interpreting the more difficult passages.

Among the students who gathered around him in the nineteen twenties and became his disciples were members of the Worker-Student Society, like Takayama Gi'ichi, Shiraishi Bon, Horie Muraichi, and Kobayashi Terutsugi; and also Miyakawa Minoru, Iwata Yoshimichi, Mizutani Chōsaburō, Fukui Takahara (Kōji), Matsukata Saburō,

Akamatsu Iomaro, and Ishikawa Kōji.[7] Kawakami guided their Marxist studies and, when necessary, even tutored them in German.[8] He welcomed them into his home, and after they graduated from the university, in many cases he helped them find employment. He did not think of himself as the traditional paternalistic father figure (*oyabun*), however; and they did not consider themselves to be his disciples (*deshi*) in the usual Japanese meaning of the word. He did not arrange their marriages, for example, nor did he take an interest in their private lives, and they, in turn, hesitated to send him marriage announcements or inform him of the births of their children. Kawakami was too busy to follow the personal fortunes of his students and, too, he felt totally inadequate as a marriage broker. Kushida had not liked the woman Kawakami selected for him, and on the one other occasion that Kawakami had served as a matchmaker the marriage had not worked out well. His relationships with students, though warm, were strictly professional, and many times throughout the decade, Kawakami found himself challenged to defend his interpretations and even his political behavior in the face of angry criticism from some of his prize pupils.[9]

Participation in the proletarian movement, even in a minimal way, as a speech maker for labor unions and a faculty adviser for Marxist study clubs, placed Kawakami in an awkward position as far as his administrative duties were concerned. The chairman of the faculty of economics as well as the mentor for a clique of radical students, Kawakami was perpetually torn between his obligations to colleagues and university officials and his desire to protect his student following. University affiliation afforded some protection from government interference, but university policy strictly prohibited faculty members from participating in outside political activities.[10] Kawakami faced censorship problems too: the fourth issue of his private journal narrowly escaped a publication ban and his article on the Russian Revolution of 1905 was censored after it appeared in the April 1921 issue of *Reconstruction*.

After about 1921 and until the end of the decade, Kawakami agreed to follow the advice of university colleagues and friends, like Kushida Tamizō, who, worried that Kawakami's career and health might be jeopardized by embroilment in political battles, persuaded him to limit his Marxist activities to classroom lectures, translations, and the monthly publication of his journal. University faculty members even dissuaded him from becoming involved in any way with the Ohara Research Institute. As a special favor to Kushida, Kawakami had asked Kōbundō Publishing Company to published the first issue of the newly reorganized institute's journal, in which Kushida's article on " 'Production' and the 'Laws of Production' in the Formula of Historical Materialism" was scheduled to appear. Kawakami hoped to establish academic ties between the university and the institute, which was dedicated to the study of social problems, but his colleagues discouraged this plan too, pleading with him to avoid even giving free talks there as he had intended to do. They eventually prevailed upon him to sever all official contacts with the research institute, though he continued to offer his help, whenever requested, to individual associates.[11] Aside from giving university lectures, speaking before campus-based study groups, and attending to administrative affairs, Kawakami withdrew after this time to the seclusion of his study.

Professional obligations left Kawakami almost no time for family life. Of necessity as well as out of personal preference, he had always remained aloof from family matters, leaving household affairs in Hide's able hands. This ability to concentrate on work, to the exclusion of all else, meant that he frequently neglected important practical and family considerations. His daughter recalled that on the few occasions when he took her and her sister on walks, he was unable to remember, if asked by friendly passers-by, how old his daughters were.[12] His brother-in-law remembered Kawakami's unwillingness to stand vigil at his son's bedside during the last days of his fatal illness. Only Hide's pleadings succeeded in prying him from his upstairs study, where he was deeply absorbed in writing an article on the inevitability of class war. Indeed, absorption in work may have been one escape from his tragic family situation. When he was not beset by immediate publication deadlines, speaking engagements, or pressing academic business of one sort or another, he frequently fell into a state of profound depression over his son's incurable heart ailment.[13]

By present-day Japanese political standards, the opinions which the authorities thought dangerous in Kawakami's writings after 1919 were far from revolutionary. Although he argued that defects in the private enterprise system necessitated changes in the social system, his specific program for reform in the years immediately following his decision to study Marxism merely called for certain basic liberal-democratic assurances, such as the right of free speech and free thought, academic freedom, the legal recognition of labor unions and strikes, and universal suffrage. It is true that he warned of imminent class war, unless workers' demands were met; and in a 1921 article he likened conditions in Japan to those in Russia at the time of the 1905 Revolution. However, despite his stirring portrayals of the Russian revolutionaries in several articles written between 1921 and 1923, he never considered the concrete act of revolution to involve him personally, nor is it likely that he expected his emotional appeals for reform to be translated into violent acts. It came as a great shock to him in 1923 to learn that Namba Daisuke, a young man inspired by his article on the heroes of the 1905 uprising in Russia, had made an attempt on the life of Crown Prince Hirohito. The assassination attempt, recalling the alleged plot by anarchists a decade earlier on the Meiji Emperor's life, was disturbing enough to bring about the resignation of the entire Yamamoto cabinet. Namba was apprehended and sentenced to death.[14]

In no sense an accomplice in the Toranomon Incident, as it was later dubbed, Kawakami was nevertheless criticized at the time for being a socialist and for having influenced Namba, who had visited Kawakami's house just before attempting the assassination.[15] But although Kawakami's name after this time became variously associated by the public with the so-called crisis thinkers at the universities and with the anarchist and socialist movements, Kawakami, as late as 1924, was a *Marukusu gakusha,* a scholar of Marxism, and not yet a *Marukusu shugisha,* a confirmed Marxist ideologue.

The Impact of Leninism

In 1925 a new contender for the crown of theoretical leadership emerged from almost total obscurity to challenge Kawakami's title as the foremost interpreter of Marxist theory. Arriving home from Germany in 1924, his steamer trunk lined with the works of Lenin, Fukumoto Kazuo (1894-) laid claim to being the carrier of the most up-to-date and authentically scientific Marxist learning. A higher school teacher by profession, Fukumoto expounded Marxist theory with so many high-flown phrases and technical terms, he managed to convince many, including Kawakami eventually, that he was a bonafide transmitter of the true faith, Russian Marxism. Fukumoto's special vocabulary soon became current jargon among enthusiastic student followers.[16] What most qualified him as a theoretical guide, however, was his alleged mastery of dialectical materialism, a mastery presumably aided by his two years in Europe studying philosophy under a private tutor.[17] His message, once deciphered, was a simple one. Following Lenin's example, he called for the unity of theory and practice, and with that end in mind, set out to revive the recently dissolved Communist party. As young students drew to his side, the newcomer rapidly succeeded in capturing the leadership of the Communist movement.

Although a spokesman for revolutionary practice, Fukumoto spoke insistently of the need for ideological clarity, a prerequisite for party strength that had been emphasized by Lenin himself. No sooner had Fukumoto returned from abroad than he chose as his primary target for attack the Kyoto University economics professor who, by this time, was exerting more influence over the minds of the nation's young students than any other Marxist theoretician.

Fukumoto challenged Kawakami on two counts: the first, that he did not understand dialectical materialism and the second, derived from the first, that he failed to appreciate the essential connection between theoretical formulation and practical activity.[18] To verify one's mental construct of the world required the practical interaction with that world. For Fukumoto, as an interpreter of Lenin, this meant that the scientific theories of Marxism had to be tested on the battleground of revolutionary practice.

Kawakami was little inclined to agree. Economics had been an academic pursuit with him for so long and objectivity and scholarly caution had become such an ingrained discipline that he hesitated to accept the Leninist theory of knowledge without careful scrutiny of it. "I do not scorn those men who throw themselves into practical movements," he explained, "but I think that by remaining outside, I can avoid the pitfalls that practitioners are prone to, and keep a scientific detachment while observing the facts."[19]

He saw that direct participation in politics was the path many of his students were following. He himself had started them on that path, almost inadvertently, and Fukumoto had pushed them further along. But these were younger men and from a different mold. Kawakami was a gentleman scholar of a past generation. Once, many years ago as a child, he had dreamed of becoming prime minister, but that dream had faded in his early manhood, and ever since he had relied mainly upon his pen to "subjugate the world."

Moreover, he was a sick man. Throughout his life he had suffered from various stomach ailments; he also had a rapid pulse and he walked with a peculiar tottering gait that added to the impression of physical weakness he gave to others.[20] Poor health had finally forced him in 1924 to give up his position as chairman of the economics department, a necessity he actually welcomed, since freedom from administrative tasks gave him more time for his own research. But one year later, he still had not fully recovered his strength. "Before the lecture I was pressed by preparations," he told Kushida, in a typical complaint, "and after the lecture, I was exhausted."[21]

Although he made strenuous efforts to popularize Marxism, the gaunt-cheeked professor with the small black moustache was no revolutionary. He was actually as little at home with the masses as Kushida and as dedicated to his research. He hardly shared Lenin's sentiment that "it is easier and more useful to experience revolution than to write about it."[22] The small study on the second floor of his home was where he felt most at ease. Here he would sit working for hours at a small desk before the window, tobacco and ink neatly arranged before him. With the exception of a daily walk through Yoshida shrine, and the company of his family and a few students and friends, he neither required nor permitted any other diversions. Should guests happen to visit on the day before a scheduled lecture, he would greet them at the front door with "I am very busy right now" and send them unceremoniously on their way.[23]

Both in temperament and intellectual training, Kawakami was ill-disposed to accept Fukumoto's assertions. The irony of Fukumoto's criticism was that it pushed Kawakami in two opposite directions at the same time, leading him high into the rarefied atmosphere of abstruse theoretical studies, while at the same time, directing him down onto the streets of revolutionary turmoil.

The Meaning of Dialectical Materialism

Kawakami was not certain first of all what was meant by this new term dialectical as opposed to historical materialism. Marx's materialist conception of history applied only to human society whereas dialectical materialism—the application of the dialectic to nature—was Engels' contribution to Marxist philosophy. It was Engels who had coined the term dialectical materialism and used it as a universal principle applicable to nature as well as to human history. By thus extending the reaches of the materialist conception of history beyond the borders of human society, Engels had tried to enhance the scientific appeal of Marxism in an age of growing positivism. His entry into the fields of science and mathematics, however, had served only to weaken the entire Marxist structure by leaving it sorely vulnerable to refutation on empirical grounds. The scientific terminology encasing these excursions into the realm of nature, moreover, had turned Marxian theorists away from the practical, political concerns of the day. The Marxist philosophical tradition was carried so far away from its original moorings that Marx is said to have commented at the end of his life that "whatever else he might be, he was certainly not a Marxist."[24]

Lenin gave the final flourish to the philosophy of dialectical materialism by emphasizing the interrelation of theory and practice, hoping to rescue, in this way, what he considered the revolutionary core of Marxist thought from the obscurantism into which it had fallen at the hands of academicians.[25] Retying theory to practice in a concrete way, he made dialectical materialism the official creed of a revolutionary party and later of the Soviet state. "Enough idle talk!" he exhorted more moderate colleagues in 1918. "We need action and action!"[26]

At the same time Lenin was determined to maintain the veracity of the Marxist world view in the face of more recent advances made in the natural sciences that were undermining faith in a material model of the universe. Against those who agonized over the "disappearance of matter," Lenin persisted in maintaining that basic reality was material, that change took place according to certain processes of inner contradiction, and that these dialectic laws of change could be understood and described by the special logic of dialectics.[27]

To show how this material reality was knowable, the dialectical materialist was forced to work out a theory of knowledge which would demonstrate how the subjective individual's knowledge of the external world could correspond with the actual structure of that world. This basic philosophical problem has of course long divided idealists from materialists, but Lenin was faced with the problem of finding a theoretical position that neither introduced the mental categories of the idealists nor fell back on the discredited mechanical materialism of the sensationalists. He attempted to resolve this difficulty by arguing that knowledge is acquired through human practice.[28] Dialectical materialism asserts that "social practice alone provides the test of the correspondence of ideal with reality, i.e., of truth."[29]

The immediate significance of Lenin's questionable contribution to the theory of knowledge was that philosophy was placed at the service of the Russian Communist party in its determination of political strategy. Disputes over epistemology in many cases merely masked what were no more than political disagreements within the Russian Marxist camp. Nevertheless, some of the disputants, such as Deborin, were sincerely interested in the purely philosophical problems involved in developing a consistent Marxist world view based on dialectical materialism.[30] Russian Marxists, generally speaking, were as concerned with epistemological questions as their German predecessors; Lenin was no exception when he urged his colleagues to read Hegel.[31] It was important to him that the law of the dialectic survive, despite the revolution in the natural sciences.

Kawakami, who had no factional bone to pick, also took the Russian Marxist debates seriously. These tortured arguments over theory represented hurdles in the way of his grasp of truth. Each discovery of an error in interpretation became a kind of epiphany, illuminating another step along the path to perfect knowledge of capitalist society, and then of the entire historical process, and finally, with dialectical materialism, of the life process itself. This knowledge guided men in the task of improving human society. "I believe that scholarship—the grasping of reality—is the basis for the increase in the welfare of mankind."[32]

Kawakami asked himself in his later years what had spurred him to such dedication and realized that the whole-hearted devotion he had given to his Marxist studies resembled the passion that had driven him into the Selfless Love movement in his student days. "With an enthusiasm that makes me wonder whether I wasn't mad, I engaged for a short while in religious movements. This was even before I put my heart and soul into the study of Marxism. That frenzied energy of the old religious movement once again was devoted whole-heartedly to Marxist research."[33]

Fukumoto's criticism challenged Kawakami's view of scholarship and demanded that he pry himself away from his ivory tower in order to put his theories to proof. The intensity of Kawakami's commitment to scholarship was not enough to qualify him as an authority on Marxism, Fukumoto argued, because the truths of Marxist theory had to be tested in the fire of revolutionary practice. This was the message of dialectical materialism.

In the face of criticism from Fukumoto, Kawakami redoubled his efforts to study German philosophy as a prerequisite for tackling dialectical materialism and responding to his arch rival. He briefly attended the lectures given by Nishida Kitarō and Tanabe Gen, both recognized authorities on German idealism, and even arranged a seminar with other members of the economics department who, like himself, were interested in Marxism and eager to expand their understanding of dialectical thought. Under the direction of Nishida and Tanabe, these discussion sessions, bringing together the representatives of two major philosophical camps, might have played an exciting role in the history of ideas in modern Japan. Instead, they passed quickly out of existence, without a true dialogue ever being established. Nishida, who held a low opinion of the materialist position, chose to limit discussion to his own philosophical preference. When a young assistant professor in the economics department introduced his own understanding of historical materialism into the discussion, Nishida commented tersely, "Superficial." The study group broke up shortly thereafter.[34]

While Kawakami struggled with German idealism in his attempt to digest the flow of new materials from the Soviet Union, Fukumoto continued to gain converts to his side. He repeatedly reminded his impressionable young audience, "We must free ourselves from Dr. Kawakami's historical materialism," because Kawakami failed to grasp the "dialectical unity of theory and practice."[35]

It is only natural that Kawakami's theoretical leadership would soon be supplanted by younger, more committed activists of a far more revolutionary persuasion. Kawakami's scholarly perseverance, which had commended him to students in the early nineteen twenties, guaranteed that he could not remain a hero to more action-oriented radicals of the latter half of the decade. Nor could these radicals accept his stubborn adherence to an old-fashioned humanist vocabulary. His goals of "hearing the Way" and of living according to the ethical imperatives of absolute unselfishness became as outmoded as his kimono in the new politics of the late Taishō period. Kawakami was slow to accept the

Marxist doctrine that politics was not a moral vocation of the elite but the struggle for power by the masses, and that the way to reform society was not to reform men's hearts but to change their economic institutions.

Kawakami himself could sense the air of excitement caused by Fukumoto's words. Growing numbers of students were participating in study groups springing up on campuses all over the country. The passage of a universal manhood suffrage law had all the more encouraged students' political efforts on behalf of the proletariat. Surrounded by this swelling radical tide, Kawakami pored over his books in a desperate effort to ascertain, before the tide engulfed him, whether or not his Marxist critics were correct. His sober life style and idealistic preachments—those very qualities which at first had drawn students to his side—were rapidly becoming anachronistic, and he was threatened with early extinction, like an overspecialized species that had failed to adapt to changing times. "Will a half-sick man addicted to reading Marx ... move society?" he wondered. "With a kind of competitive spirit, I am closeted in my study, eating from my lunchbox, working from morning to night."[36]

Kawakami's response to Fukumoto's attacks was cautious. "Fukumoto's criticism is quite powerful," he told Kushida, "but what he has written is simplistic and, I think, lacks the profundity to instruct men. I think he falls into the trap of being pedantic."[37] Unlike Kushida, Fukumoto lacked high enough social status within the intellectual world to earn Kawakami's respect: he was a mere upstart. "The Fukumoto Kazuo who frequently criticizes me in *Marxism* magazine," he wrote Kushida, "is a graduate of Tokyo Imperial University, but now he is teaching at Yamaguchi Vocational Higher School. I thought there must be two different people with the same name, but it is the same person. This is what the world has come to: Marx being done by someone teaching at a vocational higher school."[38]

At the same time, Kawakami felt uneasy about his own understanding of the complex theories of Marxism: "I am always falling into complete misunderstanding in my 'outlines' and 'popularizations.' " He was sensitive about his reputation as a mere popularizer of Marxism; and at the end of his life, he confessed that he had been nothing but a sign-carrier (*chindonya*), an advertiser for Marxism, and that Fukumoto's criticism, together with Kushida's, had been necessary to correct his own faulty scholarship:

> Since a Marxism lacking in the philosophical basis known as the law of the dialectic cannot exist, it goes without saying that the many essays concerning Marxism which I had come to write up until this

time [1925] were all very far from a true understanding of Marxism. These did nothing but scatter a vulgar, shallow explanation of Marxism and, if they had any value at all, it was to attract the interest of people to Marxism. Fortunately, I had a position as a university professor and so I had great power to influence others ... Kushida's and Fukumoto's attacks on me may have been subjectively motivated, but objectively they were pursued by social necessity.[39]

After publishing his second major attack on Kawakami, Fukumoto received an invitation to address a general meeting of the Gakuyūkai, a Kyoto University student organization. Stylishly attired in white shirt, black coat, and tails, Fukumoto arrived on his rival's home ground in late autumn 1925 to deliver his talk on "The structure of Society and the Process of Social Change."[40] Seated in the audience was the object of his severe criticism, dressed in the dark kimono that was his daily garb, and forced to listen in silence to the younger man's severe attack on his idealistic tendencies and his misunderstanding of dialectical materialism.

Kawakami confided to a friend the following day that he had not agreed with all of Fukumoto's formulations of the complicated Russian Marxist theories, but he had felt his own knowledge inadequate to make a reply. Hence he had kept silent throughout the ordeal.

An even more painful scene followed. Shortly after Fukumoto's upstaging of their hero and guide, the members of the Social Science Study Club (Shakai kagaku kenkyūkai) decided upon a formal parting of the ways. Selecting Fukumoto's various writings as their texts,[41] they moved steadily in a more radical direction, and cognizant of the breach between their teacher's scholarly disposition and their own activist inclinations, they disengaged themselves at last from his leadership. One of Kawakami's favorite students, Iwata Yoshimichi (who later joined the Communist party), was chosen to break the news to him: "A raging torrent lies before us," he said. "From now on we must link arms, leap in, and cross to the other side. We cannot expect you to lead us. You seem to have tasks to perform on this side of the bank."[42] He accepted their words without protest. On the issue of political involvement he remained undecided.

Events moved so quickly after this time, however, that Kawakami soon lost his scholar's prerogative to deliberate in leisure. A nationwide round-up of students suspected of leftwing tendencies delivered hundreds of young activists into the hands of the police. Emboldened by the new Peace Preservation Act the police scoured university campuses for suspects and entered private homes looking for subversive materials. The Kyoto campus was hardest hit. Police officials rounded up more students from that university than from any other campus. Included among those held for questioning were members of the Social Science Study Club, and even the Kawakami home was searched, although the professor himself was left undisturbed.[43]

In the midst of the furor, the university president, Araki Torasaburō, greatly alarmed and under pressure from the Ministry of Education, asked Kawakami to serve as faculty adviser for the study group, to lend an air of respectability to the students' activities. Kawakami found the task thoroughly disagreeable and was not inclined to accept it, especially since the students had already turned away from his theoretical leadership. The harassed Araki finally gained Kawakami's consent, largely because Kawakami could not bear the president's groveling. Bowing low, his bald head with it bumpy contours exposed before Kawakami's embarrassed eyes, Araki begged him to accept the position. "In all my life," Kawakami recalled in disgust, "I have never seen a man dressed in Western clothes in a Western-style room bow so low."[44]

Much against his wishes, Kawakami was dragged into the students' affairs, and when another faculty member, Watsuji Tetsurō, publicly criticized campus groups for using research as a pretext for revolutionary training, Kawakami took it upon himself to reply in their defense. Denouncing the Peace Preservation Act as bad law, he contended that the students had the right to study dialectical materialism, which, he said, they believed to be a true science.[45]

Nevertheless, Watsuji's point was not without substance. The issue involved in the January 1926 campus arrests went far beyond the guarantee of academic freedom. Student groups across the country had become infiltrated by avowed Communists. By the end of 1926 the Community party, under Fukumoto's leadership, was operating through labor organizations, the fledgling proletariat party, and student groups such as the study club on the Kyoto campus.[46] It was no longer possible to pretend that the students' right to pursue their research deserved legal protection, in cases where such research involved illegal political activity off campus.

Kawakami understood this as well as everybody else. Once he accepted Lenin's call for the unity of theory and practice, he would have to surrender the academic immunity he had worn like a shield for so many years. Yet, he completed his "Personal Settlement of Accounts With Historical Materialism" ("Yuibutsu shikan ni taishite jiko seisan") only in August of 1927. Whether intentional or not, it was in the same month that Fukumoto was read out of the Japanese Communist party.

One of the charges made by the Comintern against Fukumoto was that his over-intellectualism had cost the Communist party the support of the masses.[47] The accusation, it is true, provided the Comintern with an excuse for changing party tactics, with Fukumoto serving as a convenient scapegoat. It is also true, however, that a rigid formalism had calcified theoretical debates over Marxism in Japan. Whereas Fukumoto was not the first nor would he be the last Communist guilty of that particular tendency, he had done a great deal to exacerbate it.

This can be seen in Kawakami's final reckoning with Fukumoto over the issues that had divided them. Only those who had schooled themselves in the philosophy of dialectical materialism, as Kawakami had for the past three years, could have followed and appreciated many of the picayune points over which he labored. The long series of articles on historical materialism he published in *Social Problems* beginning in August 1927 proved that Kawakami had so well assimilated the dialectic method of thought, that he was competent to engage in what were essentially family squabbles within the fold of Russian Marxism. He and Fukumoto, for example, exerted themselves in heated exchange over the question of whether mutual, reciprocal processes necessarily excluded the law of causality.[48]

Although he still held his ground on some of the more esoteric philosophical issues (playing Deborin to Fukumoto's Lenin), Kawakami submitted to the strong influence of both Fukumoto and Kushida. He now agreed that the dialectic mode of thought was the basis of Marxist economics and that the method of the dialectic was also the basis for Marxist practice.[49] One major theme emerged from the many pages of his "Personal Settlement of Accounts with Historical Materialism": "In human praxis alone lies the criterion for the truthfulness of any theory."[50] A serious dent had been made in the professor's shield of detached objectivity.

Lessons in Class Warfare

Several months later, Kawakami took his first hesitant steps along the road to political warfare. Overcoming his dislike both of long journeys and of public speeches, he ventured back to Tokyo after eight years absence from the capital, to give a speech on "An Age of Unusual Difficulties." Here for the first time in his long career as a professor he experienced the indignity of being interrupted by the police in the middle of a talk.[51] The experience confirmed his fears: the path which lay ahead was treacherous and unmarked, but these very dangers stimulated a little of his old impetuosity, tempting him to plunge ahead.

He returned to Kyoto feeling he had already made a partial commitment to the proletarian movement. When *Central Forum (Chūō kōron)* offered to publish his banned speech, Kawakami requested the highest price possible and contributed the sum to the Labor–Farmer party. The trip to Tokyo was only a first step, and it had cost him a good deal of physical and emotional stress. Nevertheless, turning the events of the past few days over in his mind afterward, he found that "although I am gasping for breath, I want to climb the steep hill."[52]

Ōyama Ikuo was the second to chip away at Kawakami's armor of objectivity, when he asked Kawakami's help in his campaign for election to the Diet on the Labor–Farmer party (Rōdō Nōmintō) ticket. The 1928 elections, the first to be held under the Universal Manhood Suffrage Act, offered the proletarian parties their first real opportunity to gain political representation. For this reason, the election campaign promised to be a volatile one, with the left-wing movement in need of all the support it could muster.

Kawakami reluctantly agreed to join the campaign. He had known Ōyama ten years earlier, when the two men had been working for the Osaka *Asahi* newspaper. After the Osaka *Asahi* incident, Ōyama had helped found the magazine *Warera*, to which Kawakami had contributed occasional articles. It was not until 1927 that the two men came together again, this time to put out a publication called *Lectures on Marxism (Marukusushugi koza)*. By this time Ōyama, who had been instrumental in efforts to organize a proletarian political party, had become chairman of the central committee of the Labor–Farmer party.

Campaign stumping for Ōyama gave Kawakami his first direct contact with the realities of party politics in Japan. Ōyama's party was particularly vulnerable to government interference, since its membership was open to Communists. Serving as a kind of claque, Communist party members would accompany candidates on their campaign tours and, at appropriate moments in the speeches, shout defiant slogans such as "Down with the reactionary Tanaka government!" or "Give us liberty!"[53] As a result, the police kept a specially tight leash on party candidates, interfering with attempts to hold political rallies and interrupting campaign speeches when they exceeded the bounds of official tolerance. This point usually came early in the speaker's talk.

The obstructionist tactics of the authorities made Kawakami even more uncomfortable in his unaccustomed role as speech maker. However much he believed in joining theory to practice, he dreaded the task he was called on to perform. "I hate speeches and I hate traveling, but I said to myself they are unavoidable . . . and every day I stood up on the platform and each time I was ordered to stop."[54]

Ōyama lost the election, but the Labor–Farmer party succeeded in putting two other candidates in the Diet. One of them was Mizutani Chōsaburō, Kawakami's former student; the other was Yamamoto Senji, a professor of zoology from Kyoto Imperial University. The party also managed to win almost 200,000 votes, two fifths of the total vote cast for all the proletarian parties. While this figure was a mere drop in the ballot box, it represented enough of a gain to alarm the government. One month after the election, the police conducted another round-up of left-wing elements, and this time, 12,000 persons were seized, with the result that the Labor–Farmer party, its membership and support drained, was forced to disband.[55]

Following these March 25, 1928, arrests, presidents of all universities were asked to purge their own ranks of left-wing sympathizers, and students beginning the spring term at Kyoto Imperial University found that Dr. Kawakami Hajime's name was no longer on the faculty list. President Araki had accused Kawakami of improper behavior in campaigning for Ōyama and of writing improper articles for the Marxist pamphlet which he and Ōyama were putting out together. Araki had also charged Kawakami with leadership of the Social Science Study Club, even though he himself had begged Kawakami to lend his name to the group. Kawakami at first had refused to leave the university for the reasons presented, and yet, when a faculty meeting voted to ask for his resignation, he withdrew his opposition, bowing to the faculty's wishes " in the name of university self-rule," and expressing his gratitude for the years of leisurely study he had enjoyed, years in which he had traveled his painstaking road to Marxism.[56] When he left the university, students in the economics department gave him a rousing farewell party on campus and afterward accompanied him to the gate of his home near the campus, chanting "Banzai!"[57]

With surprising equanimity, Kawakami quietly gave up the teaching position he had held for twenty years. Perhaps he suddenly saw in this seemingly unjust turn of events a chance to simplify his life. Released from his teaching responsibilities and the involvement with the study club, he felt more strongly than anything else a gratifying sense of relief, as though he had "dropped a burden that is too heavy for me to bear."[58]

The last few years at the university had taken their toll on him. His son's death in 1925, though it came after years of illness, had stunned him: for a time it appeared as though he would never again resume writing. His eagerness to complete the translation of *Capital* had pulled him out of mourning, however, and, still driven by his passionate commitment to his work, he now made plans to begin translations of other basic works in the Marxist canon and to complete his simplified interpretation of *Capital*.[59]

The Kawakami family cheerfully accepted this sudden change of fortune. No longer confronted with the medical expenses incumbent upon their son's invalidism, they felt capable of meeting their financial needs even in the absence of a monthly salary from the university.[60] Royalties from Kawakami's books provided a generous income, and Hide's frugal management of their money was something Kawakami could always depend upon.[61] Besides, their expenses were lightened after the marriage of their eldest daughter. With their household reduced to three (the youngest daughter was still in high school), they accepted without alarm his enforced retirement.

Nearing his fiftieth birthday, Kawakami gave to everyone about him the impression that he was resigning not only from his professional position but also from the immediate political concerns of the day to enjoy the later years of his life in quiet fulfillment of his scholarly undertakings. Yet, the succeeding years found him more deeply enmeshed than ever in the turmoil and violence of political warfare.

10. working for the communist party

It is doubtful whether Leninism alone could have driven Kawakami from his study. After resigning from the faculty, he showed little enthusiasm for the outside world as he set out to tackle the materials on Marxism piled before him on his desk. But it had become impossible to protect his seclusion. Former students coming to visit showed him the bruises on their arms and legs where the police had beaten them. Active Communists, knowing him to be sympathetic to their cause, stopped by to borrow money.[1] He was already involved in the proletarian movement indirectly through his publication with Ōyama of the pamphlets on Marxism and his own *Research in Social Problems*. As police security tightened, he felt he had to commit himself entirely or remain aloof. True, commitment could involve anything from rousing public speeches to donations of money, and a sympathizer could be arrested for either. Already identified with the Communist cause, Kawakami faced the choice of being squeezed into a corner, keeping silent and betraying his students, or leaping into the battle to fight on the Communist side.

His friends saw another way. They believed that he had a noble task to perform in the transmission of Marxist theory and that his contribution could best be made the way it had always been made, in the translation, interpretation, and popularization of Marxism. "*Sensei,* you are mistaken," Kushida had told him, when Kawakami came to Tokyo to help form a new labor-farmer alliance. "You are not the kind of person to come out for practical movements." Kushida had been startled by Kawakami's ludicrous appearance, for he had exchanged his wooden clogs for knee-high rubber boots and his kimono for the two-piece suit he had worn over fifteen years ago in Europe. He was even more surprised by the older man's sharp query, "Don't you think that you too ought to work for the proletarian party?"[2]

It was more a criticism than a question. Kushida had already made it clear he would not join the Communist movement and had broken his resolve only once, to campaign for Ōyama on the old Labor-Farmer ticket. He had compromised then for purely personal reasons: his acquaintance with Ōyama, like Kawakami's, dated back to the days of the Osaka *Asahi* and the magazine *Warera*.

For the most part Kushida felt it was wiser for himself and Kawakami to concentrate on Marxist studies, instead of dissipating their energies in areas they were poorly suited for. Moreover, Kushida believed that the slogan of the "unity of theory and practice" which Kawakami flaunted should be understood in theoretical terms, not through what he considered to be mistaken practice. "Working shut up in your study is definitely not opposed to Marxism," he said, trying to reason with the determined man before him. "I think you, of all people, should do just that."

At these kind words of advice, which came painfully close to touching the real issue—Kawakami's internal conflict—Kawakami did what he had not done in all their years of intellectual debate. He lost his self-control. Becoming enraged, he accused the well-meaning Kushida of cowardice.[3] But his anger was really directed against himself, because he knew Kushida was right: he was not a political activist.

> The work most suited to me was the translation of *Capital;* involvement in such things as the proletarian movement did not suit me. More than anyone else, I myself best understood this.
>
> Yet, I could not rest at ease in a life absorbed in the literary work I loved best of all, closeted peacefully in my study, shutting my eyes to the movement before me. Someone engaged in translation has the right to excuse himself on those grounds. But no matter how necessary I told myself the work was, I could not be completely at ease throwing myself into it.[4]

Although political action had become the logical, though disagreeable conclusion of his search for scientific truth, allegiance to scientific truth was not the sole source of his decision to engage in revolutionary practice. Fukumoto's exhortations had been convincing, but a deeper and more powerful force than "scientific consciousness" was propelling him into action. Revolutionary practice commended itself to Kawakami precisely because it was such a difficult and unpleasant road to follow. He ironically reversed Lenin's maxim to read: "It is easier and more pleasant to write about revolution than to participate in it."[5]

In fact, the very painfulness of political activity was for him one of its chief virtues. His active participation in the proletarian movement began to take on compelling moral value precisely because it involved extreme sacrifice on his part. He began to suspect that it was morally wrong for him to stay at his desk, because that was what he really wanted to do, and therefore, he would be catering to his own self-interest. It was this curious interpretation of Leninism that made political practice unavoidable. The ethics of absolute unselfishness, which had disappeared from the pages of his Marxism, reappeared in full battle array as the basis of political practice. In the end, he left the security of his study not to correct his theoretical formulations, but to satisfy the demands of his conscience.

The New Labor–Farmer Party

Nothing but his obsession with political practice as a moral obligation could have driven him to make his trip to Tokyo in December 1928, a half year after leaving the university, against the advice and wishes of all those close to him. He had read in the newspapers about the meeting called to plan the reorganization of the Labor–Farmer party, and he had voluntarily decided to attend. Hide had accompanied him as far as the railroad station in downtown Kyoto, where she had burst into tears, pulling at his sleeve and begging him not to go. Here again, as with Kushida, he had turned his own fears into anger, lashing out at her with harsh words: "What's this nonsense you're saying!" Yet his own reluctance was all too real. "I myself did not want to leave. I was in a quandary. I did not want to go but there was a sense of obligation that I should go. That's what pushed me. I was pushed, pushed by that force and finally, with a burst of effort, I left ... actually feeling wretched."[6]

As Hide had feared, the police broke up the gathering, using the opportunity to detain her husband, together with all the provincial delegates to the meeting. This initial brush with the law, however, only strengthened her husband's will to resistance. He returned from his ill-fated Tokyo adventure more determined than ever to work for the proletarian movement. He even allowed himself a brief moment of heroic exuberance. "University professors spend their winters around the *kotatsu* (footwarmer)," he wrote, shortly after the incident, "but if once they stir themselves and try going out, pleasant scenery opens before them which those who spend the year clinging to the *kotatsu* cannot see even in a dream."[7]

News of the half day he spent in the jailhouse made the headlines of that evening's Tokyo newspapers. The arrest also brought the name of the former economics professor and Marxist translator to the attention of the Russian Communist world. For many years afterward, Kawakami saved the letter he received (in English) from Ryazanov, head of the Soviet Union's Marx–Engels Research Institute, who asked about the progress of his translation of *Capital* and declared himself "ready and most willing to give all the aid in my power to the well-tried Japanese Marxists."[8]

Kawakami's first step upon returning to Kyoto was to announce that his own magazine, *Research in Social Problems,* would thereafter become a magazine of the proletariat.[9] He placed its publication in the hands of Suzuki Yasuzō, a former graduate student in economics at Kyoto Imperial University active in radical campus movements, while he turned all his attention in the two years that followed to the task of rebuilding the defunct Labor–Farmer party.

Having at last made up his mind to join the proletarian movement, Kawakami plunged in with the characteristic wholeheartedness that had become a hallmark of his personality. There were no shades of gray in his color wheel of values. Once he had fixed his sights on a certain goal, he went after it with an intensity which may have seemed fanatic but to him signaled his absolute sincerity of purpose. One is often uncertain which was more important to Kawakami, the goal itself or the unstinting manner in which he pursued it. Perhaps the two were indistinguishable. "The believer in an ethic of ultimate ends," Max Weber has remarked, "feels 'responsible' only for seeing to it that the flame of pure intention is not quenched."[10]

Whenever something appeared before his eyes as the truth, Kawakami wrote in his "Self-Portrait,"

> ... no matter what sort of thing it is, once having accepted it, I hang on with a firm grip, and I keep digging until I have understood it. As long as it remains the truth for me, I will submit to it with single-mindedness of purpose and a humble heart—unconditionally, absolutely, thoroughly—with regard neither for status nor reward. In the end, even if I have pushed myself into some undreamed of dangerous or imprudent circumstance, I do not flee or hang back, but responding to the highest command, I leap without hesitation over the abyss.
>
> At the same time, while marching to the truth with such total determination, should I realize that what I thought at first was the truth, is not, at that instant, I immediately throw it away, resolutely burying the past. This is the kind of man I am.[11]

Thus, the passion which he had devoted at first to the Selfless Love movement and then to scholarship, he now devoted to the dangerous and difficult mission of reviving the Labor–Farmer party.

Kawakami's efforts to rebuild a legal political party for the extreme left were based on his belief that it could pave the way for the eventual reconstruction of the Communist party. The Communists had yet to recover from the mass arrests of March 1928, and since that time members of the Labor–Farmer group had debated among themselves whether to rebuild a new legal party or wait for the Communist party to rebuild its own forces. In the interim, an underground Labor–

Farmer Federation for Securing Political Liberties (*Seijiteki jiyū rōnō dōmei*) had been created to keep Communist and pro-Communist Marxists together.

The objection raised by the opponents of a new party was that legality would severely restrict the activities of party members, resulting in the inevitable "sell-out" of the workers. The Comintern's decision to oppose the new party at that time brought the issue to a head. The Comintern-directed Communist party had already lost its original founders after the publication of the 1927 Theses. Disagreement over the Comintern strategy for revolution in Japan had led Yamakawa Hitoshi, Arahata Kanson, and Sakai Toshihiko to form their own left-wing Labor–Farmer faction, divorced from international control. Now the remaining elements of the pro-Comintern Labor–Farmer movement split again, and again over the issue of strategy. Mizutani Chōsaburō, a successful candidate for the Diet on the old Labor–Farmer ticket, chose to form his own legal party; the Labor–Farmer Federation decided to follow the Comintern directive; and Ōyama, Kawakami, and Hososako Kanemitsu, the secretary for the old party, favored pushing ahead with plans to establish a new Labor–Farmer party.[12] For their efforts, all three were expelled from the federation.

Kawakami was at this point joined in his political activities by his wife's younger brother who, under the influence of Kawakami and Ōyama, had been drawn to the socialist movement.[13] Ōtsuka Yūshō, a thirty-year-old bank clerk in Osaka, took it upon himself to assist and protect his less worldly brother-in-law in the increasingly more perilous missions into which the older man was soon dragged.

He was not at his brother-in-law's side, however, when Kawakami was arrested for the second time in February 1929 at a meeting in Tokyo of the All-Nation Farmers' Union (Zenkoku nōmin kumiai). Invited to give a congratulatory address to the group, Kawakami was interrupted by police in the middle of his remarks and taken in a motorcycle sidecar to the nearest police station for questioning.[14] The meeting of the Farmers' Union was quickly adjourned, and all those in attendance marched to the police station to demand the prisoner's release. In the midst of the uproar, as armed police tried to disband the crowd, Diet representative Yamamoto Senji, with two other men, pushed his way through the crowd and entered the building through the rear door. Despite his proletarian party connections, Yamamoto, a zoologist, still enjoyed a considerable degree of influence, and before long he emerged from the police station with the prisoner in tow.[15]

Less than three weeks later, on March 5, 1929, Yamamoto was stabbed to death in his hotel room in Tokyo by a member of a right-wing organization. To Kawakami fell the dolorous honor of presenting the funeral oration. Once again he set out by train for Tokyo, where Ōtsuka met him at the station and accompanied him to the funeral ceremony, held at a downtown theater hall. Dressed in a shabby coat, his face pale and drawn, Kawakami took his place at the front of the hall and began to deliver the oration.[16]

Seated in the audience were Ōyama, Hososako, and other active leaders of the proletarian movement, straining in their seats to hear the soft-spoken professor's words. He had prepared a moving tribute to Yamamoto, "a great scientist ... but an even greater human being," but no sooner did he begin his talk than the police inspector stationed in the hall, taking exception to his mention of "the firm will to fight," ordered him to discontinue his speech.[17]

The funeral service instantly turned into wild pandemonium, as the audience began chanting, "Oppression by government officials!" In the melee that followed, the police inspector climbed onto the stage to try to quiet the chanters and was pushed off by one of the members of the audience. Remarkably enough, the party leaders managed to escape arrest.[18]

Kawakami did not allow these newly gained experiences of violence and police suppression to deter him from his chosen path. The martyred Yamamoto Senji served as fresh inspiration, for he was a figure with whom Kawakami could sympathetically identify: a politically committed scholar. Nevertheless his new career as a politician remained for him as strange and unreal as life in a foreign country, especially after the years of seclusion he had known. The old, ill-fitting Western clothes he donned on his speech-making excursions added to the element of the bizarre and became an object of gentle ridicule to his friends. Kushida once remarked to Ōyama, "Whenever I see the professor's boots, I feel depressed."[19] The boots, old and inappropriate, became an apt symbol for Kawakami's antiquated Marxism.

Wherever he went he was accompanied by the personal bodyguard assigned to him by a Kyoto labor union group after Yamamoto Senji's assassination.[20] To a man who, as Kawakami put it, had always been too absorbed in study to see the beauty of nearby Ohara in autumn, the turmoil of the campaign trail was nothing short of overwhelming.[21] At one point he became so confused by the pace of his new life that, standing in front of Tokyo University, he began looking for a restaurant located near the Kyoto campus. He had forgotten where he was.[22]

His work for the party placed a terrible strain on the immediate members of his family. Whenever he left the house, they would lapse into a state of anxious waiting. They were always worrying about his health, an index of which were his eating habits. Although he was five feet seven inches tall, he weighed only around one hundred pounds.[23] Ōtsuka records in his own memoirs how relieved he felt when his brother-in-law, arriving tired and pale to attend funeral services for Yamamoto, was nevertheless able to eat lunch.[24] Hide's greatest pleasure in those years, Kawakami remembered afterward, was to provide the most nutritious meals, in the shortest possible time, for him, his hurried visitors, and his bodyguard.[25]

Even after a second great wave of arrests in April 1929, when the extreme left of the proletarian movement seemed to have ground to a complete halt, Kawakami refused to abandon hope for the party. He wrote a letter to Hososako several months after the arrests, urging him to renew his flagging efforts, and learning that Hososako's father was opposed to his son's political involvement, he made a special trip to Yamaguchi Prefecture to try to persuade the old man. Finally, by November of the same year, the New Labor–Farmer party (Shin Rōnōtō) was officially established, with Ōyama as Central Committee chairman; Hososako, secretary; and Kawakami, editor as well as sole financial supporter of the party newspapers.[26]

The new party had little more than three months to prepare for the coming 1930 general elections. As the spark plug for its regeneration, Kawakami bore considerable responsibility for the party's successful operation. One month before the February elections, he moved to Tokyo to work in close cooperation with Ōyama and Hososako. As a final gesture of dedication, he acceded to the party's decision that he enter the elections as a candidate from Kyoto's Fifth Ward district.

The campaign itself aroused bitter feelings between himself and his former student, Mizutani Chōsaburō, who had also chosen to run for election in Kyoto, representing his newly formed All-Japan Farmer–Labor Mass party (Zenkoku rōnō taishūtō).[27] With the left-wing vote split between the two proletarian party candidates, neither one gained a seat in the Diet, although as far as Kawakami was concerned, his own defeat came as something of a relief. When the results of the election were announced, he turned to a weeping campaign worker and said, "I'm really better off having lost."[28] His wife concurred.

The party's campaign in Tokyo was more successful, putting Ōyama into the Diet. Kawakami's enthusiasm, however, soon began to wane, as he realized that Ōyama had no intention of eventually turning over his political following to the Communist party. "Deep down inside Ōyama was not a Communist," Kawakami wrote, recalling the disillusionment he had felt when he discovered that his political ally wanted to keep control of the party from falling into Communist hands.[29] "Mr. Ōyama likes speech-making better than eating or anything else. At street corner speeches, the bigger the crowd the higher his spirits. He doesn't worry about anything else."[30]

From the very beginning, Kawakami had considered the Labor–Farmer party no more than a temporary expedient to help pave the way for the official Japanese Communist party. Ōyama's reformism, he argued, was only playing into the hands of the opposition.[31] When the Comintern called for the immediate re-establishment of the Communist party, the feud between the two leaders erupted into open antagonism. Together with Hososako, Kawakami proposed the immediate dissolution of the New Labor–Farmer party, so that members could give their full support to the Communists. They left Ōyama in the fall of 1930, and the New Labor–Farmer party folded shortly thereafter.[32]

The movement to reform the new proletarian party had been a disaster. Kawakami's break with Ōyama led him into "one of the most despairing periods of my whole career."[33] He had alienated a score of friends over the issue of whether to found the party and again over whether to dissolve it. The original labor-farmer alliance was now split half a dozen ways, but as far as Kawakami was concerned, he wanted no part of any of them. Considerably sobered by the experience, he decided to retire from politics and return to his Marxist scholarship. He had done his small part, satisfied conscience and curiosity, and was now prepared to close himself off from the world to do the work he enjoyed most of all. "After wandering in a foreign country for a while, it's good to return home."[34]

Before him on his desk lay the unfinished manuscript of *Capital*, which he had begun translating three years earlier with another one of his former students, Miyakawa Minoru. Although he had managed to publish his *Brief Explanation of Capital (Shihonron ryakkai)* and to work on a simplified text of *Capital (Shihonron nyūmon)*, he had not yet finished his definitive translation of the Marxist scripture. Turning now to that task, he assured his worried family that "it is my intention to spend the remaining years of my life in my study completely cut off from contacts." This assurance, written in a letter to his son-in-law, told of Kawakami's newly gained "confidence in the work I do in my study . . . I think this time I can settle down."[35]

The End of the Road

Although Kawakami viewed his entrance into the Communist party in 1932 as the culmination of his lifetime quest, the facts, as Kawakami himself reveals them, suggest that he was pushed—or that he fell—into the party with a sense of weary resignation. Toward the end of his days, imprisoned for his Communist party activities, Kawakami tried to make the many pieces of his life fit together like the plot of a well-ordered novel. His autobiography describes his hard labor along the road to Marxism and his triumphant entrance into the party at the end, but the narrative provides merely a literary unity that does injustice to the facts.

When Kawakami was lured from the sanctuary of his own private study to work for the Communist party, he had already made his peace with Marxism–Leninism. On the level of theory, he expressed new confidence in the scientific maturity of his scholarship. "Looking back at what I wrote in the first *Tale of Poverty*," he had to "force a smile" at the person he was then. "When I see that past self, drowning in such an ethical, religious utopia, I feel that I am confronting a completely different person."[36] He now admitted that it was "impossible properly to understand Marxist economics if it is separated from its philosophical basis"[37] and demonstrated his understanding of that basis in his long and elaborate explanations of historical materialism and the law of the dialectic. His final resistance shattered, he conceded that "social existence determines social consciousness": all thought was a reflection of economic class circumstances.[38] Finally, he accepted the veracity of the labor theory of value, describing with certainty the inevitable downfall of the capitalist system.[39]

The studies he published after 1925 were accordingly free of references to eternal truth or absolute unselfishness. As his study of the history of economics was gradually transformed into the materialist conception of history, he also omitted from his exegeses biographies of Carlyle and Ruskin and the moral teachings of Smart. He believed that, after Kushida's criticism of his *Historical Development of Capitalist Economics* in 1924, he had finally become a "true Marxist." "For a long time I was unable to arrive at a dialectical understanding of the unity and opposition of religious and scientific truth," he confessed, somewhat cryptically.

> For this reason, while engaging in the scientific study of social phenomena, I would unconsciously commit such basic errors as confusing scientific truth with religious truth, mixing up the world of spirit with the world of matter, and introducing metaphysical idealism into the scientific world. When doing research, I always fell into hopeless confusion. In 1924, after publishing *The Historical Development of Capitalist Economics*, I first began to set to rights my "twenty-year-old theme of reflection upon selfish activity." By this I mean that I finally separated religion from the world of science. I had long resolved to pull myself away from religious truth which, since the time I turned twenty years of age I could neither forget nor separate from. Truthfully speaking, to me—I was already forty-six years old—it was not a twenty-year but rather almost a thirty-year long metamorphosis . . . Breaking out of the chrysalis, I became a butterfly.[40]

The one point he refused to concede was the morality of the proletarian class. He wrestled with the problem of the selfishness of the working class but finally agreed that, although they were obviously fighting for their own class interest, after the defeat of the capitalists and in the new classless society, the workers' interests would expand to include the common welfare of all members of society.[41]

On the level of praxis, Kawakami seemed prepared to interpret that word in its broader meaning to include the contribution made by the scholar's pen. He had before him as encouragement a letter written by Sano Manabu in prison urging him to stay out of politics and to devote all his energies to *Capital*.[42] He seemed to have every intention of following this advice.

By the time the first volume of *Capital* was published in 1931, however, Kawakami was too involved in one way or the other with the Communist cause to remove himself completely from the political scene. Kawakami Yoshiko, following the examples of her father and uncle, had moved to Osaka to work in the women's labor movement,[43] and Ōtsuka, who had left the New Labor–Farmer party together with Kawakami the year before, was now a member of the finance committee of the Communist party. Thus, Kawakami could not avoid the day by day political struggle going on before his eyes.

To be sure, his sense of obligation to aid the new Communist party gave him little rest. Guilt for the mistakes he had committed in aligning with Ōyama troubled him; he felt he had succeeded only in weakening the Communist movement. It was this nagging sense of obligation on his part and a desperate need for money on its part that at last brought the Marxist scholar and the Communist party together.

The connection was made through Ōtsuka in utmost secrecy. A representative of the party approached Kawakami in the spring of 1931, requesting a fixed monthly contribution for party activities. Kawakami willingly agreed to pay. He hoped in this way to discharge his obligation to the revolutionary cause.[44] In any case, with Ōtsuka on the finance committee, it is difficult to see how he could have refused.

After the arrests of 1928 and 1929 and its own private purge, the Communist party consisted of no more than a "small group of desperadoes," radical revolutionaries loyal to the Comintern and bent on achieving revolution at all costs. Its chances for success were virtually hopeless. Whether Kawakami understood this at the time is hard to say. He did know the danger he was exposed to by supporting the party with funds, since only a year before, twenty intellectuals, among them college professors, had been seized for making contributions to the party.[45]

As the position of the revolutionaries became more precarious, they pressed Kawakami to increase his monthly contributions. He, in turn, pressed his publisher, Iwanami, to increase the royalties from *Capital*, so that he could meet the party's demands.[46] Kawakami and his wife began to live in a state of constant fear that the officials would notice their large monthly bank withdrawals.[47] Almost despite himself, Kawakami seemed to be getting sucked into the revolutionary struggle.

Even in this atmosphere of growing tension, Kawakami managed to work long hours on his translations. In the early summer of 1932, however, he was suddenly contacted directly by the central committee of the Communist party and asked to translate the new Comintern *Theses*. Through a roundabout course, the Comintern's new strategy for the Communist party had arrived, in German, plucked from the pages of a magazine and sent from Berlin by a Japanese party sympathizer.[48] The party anxiously waited while Kawakami, working at a furious rate, translated the lengthy document in less than two weeks time. The *Theses* appeared under Kawakami's assumed name, Honda Kōzō, in a special July issue of the party organ, *Red Flag (Akahata)*. Translations of the *Thesis* were also smuggled into the prison cells of the arrested Party leaders. To the Central Committee, Kawakami became a minor hero.[49]

An Indian summer of vigorous party activities followed the new Comintern directive. The new strategy served to strengthen the hand of the Japanese Communists, because its reversion to a tactic of support for a bourgeois-democratic revolution enabled the party to broaden its political base. Operating through front organizations, the Executive Committee succeeded in developing a complex party machine, with branches reaching out into numerous non-Communist, left-wing groups.

As hopes for the party revived, the danger to its supporters increased proportionately. The Home Ministry, alarmed at the proliferation of "dangerous thoughts," intensified its efforts to track down Communist party leaders, suppress their publications, and stamp out their loyal following.[50]

On August 8, 1932, after a full day spent working on *Capital*, Kawakami was seated at the dinner table eating his evening meal when his brother-in-law burst into the house with the alarming news of a new wave of police raids planned to round up party supporters. This time they would surely seize the former economics professor. He had no choice but to go into hiding.

Kawakami at first refused to go, so reluctant was he to leave his familiar surroundings. Prodded by Ōtsuka, he finally decided to resist arrest, and gathering together whatever books he could carry with him, took temporary refuge in the home of a distant acquaintance. There, on the following day, under these forlorn circumstances, he learned that he had been recommended for membership in the Communist party.[51]

In the next four months, Kawakami moved half a dozen times to various hiding places in the Tokyo area provided for him by the party. His official assignment was editing *Red Flag* and other party pamphlets, a task hampered by the lack of those two ingredients most essential to his life, research materials and the assuring familiarity of his own study. Too well known to show his face on the city streets, he was totally dependent upon Ōtsuka and Hide for all his needs and for news of the outside world, hardly daring to poke his head out the window to view the first signs of autumn.

Ōtsuka served as the link between Kawakami and the party. He led party officials to Kawakami's hideouts and arranged for the moves from one house to the next. He also took pains to make Kawakami as comfortable as possible in each new location, buying a desk when it was needed or bringing in a brazier for the winter months. But it was Kawakami's wife who, as always, continued to be his greatest asset. Prudent when he was impulsive, practical where he was visionary, cautious where he was trusting, Hide was his opposite in every way except one: with him she shared that strength and tenacity in the face of opposition derived from their common background. She credited these qualities to her Chōshū upbringing.[52]

Although Hide had opposed her husband's political involvement from the very beginning, she had helped him in every way she knew how. As he rushed to finish work on *Capital*, she had sat at his side writing out clean copies of his translations page by page. She delivered clothes and books to him when he was in hiding, carrying the works of Lenin with her in a small suitcase to avoid detection. When the party asked Kawakami for more money, it was his wife who went to the bank to draw out their savings.[53] She managed to keep the household going, even though her husband, who was no longer receiving a salary, was donating royalties from his books to the party. "We are country folk," she explained, "accustomed to getting along on little money." She buried a son, married off a daughter, and succored a husband and brother through the nightmare of the leftist persecutions of the early Shōwa period, and despite all the horrors that were yet to come, she would remember at the end of her life that the worst time for her had been Hajime's two months in the Garden of Selflessness, when she had waited in Yamaguchi, wondering whether he would ever return to society. To the end of her days she insisted that if she had been with him then to take care of him, he never would have joined the religious sect.[54]

Assured of faithful backing from his family, Kawakami left the practical details of his daily life to others, as he always had, and concentrated all his attention on his work for the party. He had utmost respect for the authority of the Communist party and, behind it, the Comintern. He accepted the party's will without question, even though he had no idea who its leaders were. A young man who went by the name of Hiramatsu visited him four or five times during this period in connection with party publications, serving as Kawakami's direct link with the party. The young man turned out to be one of the key figures in the Central Committee, Kazama Jokichi. When Kawakami became aware of this afterward, he wrote of what a great honor it was for him to have been visited by such an important party official.[55]

This ignorance of party leadership, though an unavoidable aspect of the underground revolutionary organization, led to a tragic misunderstanding. In this period of changing allegiances and secret liaisons, Kawakami somehow conceived the idea that Iwata Yoshimichi, at one time his favorite student, was a government spy. Iwata was actually one of the leaders of the party, a fact Kawakami learned only belatedly, when he read of Iwata's arrest and subsequent death in prison at the hands of his captors.[56] As a final touch of irony, the third leading member of the Central Committee, Matsumura Noboru, was in fact a government agent, but here too, Kawakami learned this fact too late to make use of it.[57]

He first met Matsumura at the beginning of October. Kawakami had grown somewhat concerned when Kazama, alias Hiramatsu, had failed to show up for his regular consultations over the party newspaper. The young man he had sent to take his place explained that Hiramatsu was too busy to come himself. Then, Kazama's young delegate stopped coming, and Ōtsuka, too, remained away. Kawakami waited. Finally, on October 10, Ōtsuka arrived with two other party members, one of whom was the agent, Matsumura. Crowding into Kawakami's small bedroom, they briefly explained, in whispers, what had happened.

Ōtsuka had hatched a plot to procure desperately needed funds for the party by staging a daring bank robbery. He and an accomplice had held up the main branch of the Kawasaki Bank in Ōmori, on October 6, getting away with a large sum of money for the party coffers. What Ōtsuka avoided mentioning was that he had used Kawakami's younger daughter, Yoshiko, as camouflage, having her drive alongside him in the getaway car to lend an air of respectability to their group escape. Fortunately, Yoshiko and Ōtsuka had succeeded in outwitting the police, but the spectacular bank robbery, dubbed the Ōmori Gang affair, had put a high price on Ōtsuka's head and further endangered the lives of the remaining party members.

It was because of this new danger that Ōtsuka decided to discontinue his visits to Kawakami, and now Matsumura remained the sole connection between Kawakami and the party. Nobody, least of all Kawakami, suspected their fellow party member of treachery, and Matsumura had taken special pains to allay any doubts Kawakami might have had, by entrusting to him secret party documents, and in a moment of tense drama, demonstrating how to burn the documents should the need arise.[58] Thus, even when Matsumura miraculously avoided capture at the end of October, in the arrests that had snared Kazama and Iwata, Kawakami and Ōtsuka maintained their trust in their party comrade.

On December 1, one thousand suspects were rounded up by the police, but Ōtsuka remained free. Kawakami decided to move to a new location in the home of an artist in the Nakano Ward in Tokyo. By this time, all contact with the party had ceased. Kawakami knew only that three men had been arrested; Kazama, the man who had replaced Kazama as liaison, and Iwata; but he never doubted that some method would be found to reestablish contact between himself and those in command. In the meantime, he waited in his comfortable but lonely surroundings on the second floor of the Nakano house, putting out one copy after another of the party pamphlet, and hoping every day for some word from his comrades.

Thus abandoned, he celebrated the New Year in his generous host's home, cheered by a surprise visit from Ōtsuka, who had nevertheless little information to contribute, since he himself was trying to make contact with the party. Four days later, on January 5, while walking away from a rendezvous arranged by Matsumura, Ōtsuka was seized by the police.[59]

Upon learning Matsumura's true identity as an agent of the authorities, Ōtsuka realized it would be pointless for Kawakami to continue his underground existence. He offered to inform the police of his brother-in-law's whereabouts if they, in turn, would give him a few minutes alone with Kawakami before the arrest, so that he could break the news gently to him. Concerned about Kawakami's weak health, he feared the shock would place too much of a strain on the older man's heart.[60]

The police agreed only to carry a note from Ōtsuka to Kawakami. When they entered his second-floor room and thrust the note into his hands, Kawakami, far from being disturbed, quietly put on his glasses, read Ōtsuka's note, and said softly, "I understand. I'll go at any time."[61] Left behind him on his desk were the party pamphlets he had written, but which nobody had come to read.

On January 8, 1933, a special edition of the Tokyo *Asahi* newspaper announced on its front page the arrest of Dr. Kawakami Hajime, described in gratifyingly inaccurate terms as a theoretical leader of the

Communist party. The truth is that by then there was no Communist party, and the one hundred twenty-two days of Kawakami's underground existence had been a pathetic and absurd waste.

Yet, Kawakami treasured his membership in the Japanese Communist party and his short period of activism as the crowning achievement of his long life, refusing even in prison to denounce his party association. Looking back on his life, he viewed this final stage of his career as the culmination of many long years of endeavoring to become a genuine Marxist.

> Among the party members at that time, there were some inconsequential people. One might wonder why I was so pleased to become one of them. Let me explain why. When the Chinese Red Army moved from Jui Chin to Yenan, marching on foot six thousand miles, Hsieh Rou Ts'ai, who was already fifty, managed to keep up with the younger people in crossing the greatest river and the highest mountains in China. Even this strong man Hsieh, when asked to become a member of the Communist party, "wept with pleasure over the thought that even an old man like myself could be of some use in the construction of a new world." I am not a strong man like Hsieh. Far from it. However, I wept like Hsieh, when I was given an opportunity to join the party when I was even older than Hsieh. I, the Marxist scholar until then, was able to transform myself into a full-fledged Marxist. It was not at all easy for me ... to reach that point.[62]

News of his acceptance had come to him as a complete surprise, he related in his memoirs, filling him with inexpressible gratitude and endless joy.[63] Suddenly the burden of anxiety had lifted and, feeling he had at last accomplished his life quest, he was inspired to write a poem at the time commemorating his entrance into the party by comparing himself to the heroic survivors of the Chinese Long March: "Here, standing at my destination and looking back—/How far I have traveled across the rivers and mountains!"[64]

Are these no more than the words of a tired, old man, in jail, trying to comfort himself in his last years with the assurance that his life course had led him to a meaningful goal? Or, conversely, did the last one hundred days of his "road to Marxism," for all their outward futility, unexpectedly bestow upon him the opportunity to piece together to his own satisfaction all the loose ends of his mental drama?

All the conflicting forces at play in Kawakami's personality found their meeting point in the Japanese Communist party. As the purest embodiment of Marxism–Leninism, the party represented the joining of scientific consciousness with revolutionary practice. Through his participation in it, Kawakami could complete his long and torturous metamorphosis into a true Marxist, while at the same time give himself in an act of total self-denial for a public cause. Stamped with the seal of authenticity as a bona fide Marxist in the service of the Communist party, he could weld theory to practice and at the same time act out his lifelong commitment to an ethics of absolute unselfishness.

For what continued to motivate Kawakami in his daily life could not be expressed through the formal vocabulary of Marxism. He had never dreamt of subjecting his values to class analysis, nor of identifying his moral imperative with the ideology of a particular class. In his private being, he had never given up his belief in an unchanging, absolute religious truth of self-denial, nor doubted the reality of the irrational, unscientific process of intuition by which he had come to that truth.

Only publicly had Kawakami made his peace with Marxism; subjectively he was still ensnared by what Marx and Lenin termed the myths of the bourgeoisie. But perhaps this is because both Marx and Lenin were heartier souls. Marx never understood the inner conflicts that tortured his more romantic contemporaries—men like Carlyle and Coleridge. His mind, writes Isaiah Berlin, "was made of stronger and cruder texture; he was insensitive, self-confident, and strong-willed; the causes of his unhappiness lay wholly outside him . . . his inner life was tranquil, uncomplicated and secure."[65] Given this certainty, Marx "looked upon moral or emotional suffering, and spiritual crises, as so much bourgeois self-indulgence . . . like Lenin after him, he had nothing but contempt for those who, during the heat of battle, . . . were preoccupied with the state of their own souls."[66]

Intensely preoccupied with the state of his own soul, Kawakami could not completely reconcile, in theory, his "unchanging religious truth" with the doctrine of historical materialism. The reconciliation had to come at last as a confluence of two forces: his personal morality and the Marxist political movement. In his final sacrifice of self, he joined the materialist philosophy of Marxism with his own altruistic ethics. The Japanese Communist party came to represent for Kawakami the point of contact of the two truths he had pursued simultaneously throughout his life, moral truth and scientific truth. Hence, what begins in his autobiography as an explanation of how he "entered the gates of Marxism" ends with the description of his search for the Way. The story of Japan's social revolution takes second place to the story of Kawakami's spiritual and intellectual revolution.

epilogue – prison years and beyond

A favorite question for Japanese scholars interested in the life and thought of Kawakami Hajime is why he did not recant (*tenkō*) during his prison years. Kawakami's staunch refusal to renounce Marxism contrasts sharply with the compromises and conversions made by other Japanese Communists. Although for many years he had agonized over the validity of the Marxist world view, he held his ground in jail against his captors' attempts to convert him.

It is important, however, to define what is meant by conversion. Even the recanting of Nabeyama Sadachika and Sano Manabu in 1933, though admittedly a great shock to the remaining free Communists, did not represent total repudiations of Communist ideology. What these men renounced was international communism; that is, the outside interference of the Comintern, but not Communist revolution within their own nation's borders.[1]

Kawakami refused to make even this concession, although officials did succeed in extracting from him the pledge that he would never again resume his political activities. As one might imagine, Kawakami had little difficulty acceding to this demand: he no longer had the strength to rejoin the revolution, even if it were to begin outside his study window.

Because he agreed to refrain from political activity, however, Kawakami may deserve the label of partial recanter (*juntenkōsha*), the fifth of six officially designated degrees of *tenkō* which the Ministry of Justice applied to Marxists who, while refusing to renounce their ideology, agreed not to continue political activity.[2] Indeed, on the day he was released from prison, four months short of his fifty-eighth birthday, on June 15, 1937, Kawakami placed a brief announcement in the newspapers in order to publicly avow his renunciation of political practice. "I am using the occasion of my release from prison," he wrote, "to close my career as a Marxist. This note is that career's elegy. It is an epitaph."[3]

This time, moreover, Kawakami faithfully lived up to his promise to retire from public affairs. In the final decade of his life, he carefully guarded his privacy. Setting aside political and professional obligations, he used his remaining years to indulge in the pleasure of his grandchildren's company and in the literary and artistic joys of his youth—

writing poetry, practicing calligraphy, and completing his autobiography.[4] Although some of his Marxist works were republished by the Japanese Communist party after World War II, when party leaders, released from jail, deemed his work indispensable to the furtherance of the mass movement, for the most part, Kawakami's days as a Marxist were over.[5]

Nevertheless, Kawakami left the movement because he realized he was unsuited for it, not because he found the movement itself inadequate. He never became disillusioned with the theory or practice of communism.[6] Nor did he use his pledge to refrain from political activity as a way of buying a shorter prison term. Except for a brief period directly after he was sentenced, Kawakami stood firm in his allegiance to communism, knowing full well that in so doing, he forfeited the possible reward of a reduced sentence. Accused of being a leader of the Communist party, in violation of the Peace Preservation Act, he was tried in early August 1933, found guilty, and sentenced to five years in prison.[7] The verdict stunned him. His lawyers had argued that Kawakami had been merely a participant in the party—an offense liable for only two years imprisonment. This was the sentence meted out to Kawakami's brother-in-law, Ōtsuka Yūshō. In despair, Kawakami wrote to the court promising to give up his Marxist studies but soon afterward felt ashamed that he had lost control of himself and retracted his pledge.[8] The eventual commutation of his prison sentence to four years apparently may be explained by a special decree issued to honor the birth of crown prince Akihito.[9]

The most important reason for Kawakami's unwillingness to renounce his ideological convictions was his faith in the scientific truth of Marxism. His path to Marxism was scattered with the debris of theoretical debates. For all the emotional energy invested in his work, Kawakami was dedicated to the methodical study of objective social truth. Before accepting his critics' interpretation of Marxism, he had burrowed through all the documents and digested all the arguments. After such a painstaking process of investigation, it was unlikely that he could be convinced to change his mind on other than intellectual grounds. Without the tools of his trade before him in the form of the printed word, he was not prepared to surrender his allegiance to Marxism. "Even if I were boiled in oil, I would never depart from this academic faith."[10]

Kawakami had an almost religious attachment to research. He once confided, "I rely more strongly on my scholarship than on myself."[11] It is worth while to examine the origins of this reverential attitude toward scholarship, because the power of Marxist thought over the minds of intellectuals like Kawakami Hajime was closely bound up with their struggles to master the methodology of the modern social sciences.

In his early approach to economic study, Kawakami's writings reflected the lingering influence of Tokugawa neo-Confucian statecraft. The purpose of learning, according to the Confucians, was moral cultivation—the necessary prerequisite for public service. Study had a practical end: to improve one's own character in order to improve society. Implicit in traditional scholarship was the assumption that the morally refined man could correct social ills by the sheer power of his personal example. This is why solutions to social problems were often couched in ethical terms, and even specific practical policy proposals were accompanied by moral injunctions.

The demands of the modern social sciences for a value-free body of knowledge clashed with this older view of the nature and purpose of knowledge. Teaching and preaching, long considered indivisible, were forced apart: problems of methodology, research, and analysis involved in establishing an objective body of knowledge challenged Japanese intellectuals like Kawakami, who had been accustomed to spicing their analyses of social problems with moral platitudes and winning their technical arguments on literary points.

The introduction of the German historical or social policy school promoted a respect for historical materials, statistics, and accurate documentation, but it also reintroduced policy concerns into the burgeoning science of economics, thereby keeping alive that blend of ethics and economics characteristic of Tokugawa economic thought.

Academic Marxism finally separated ethics and economics. Its insistence on disciplined thought created enormous tensions in thinkers like Kawakami Hajime who were raised in a more holistic intellectual tradition. Convinced that the acquisition of knowledge should be related to the transmission of wisdom, he could not separate his mental life into two separate compartments, one for his professional work and the other for his spiritual ideals. His early difficulties as an economics professor stemmed from the fact that his professional career was anchored to the economics of self-interest, whereas his personal belief was grounded in the ethics of selflessness.

The systematic nature of academic Marxism was demanding for other reasons as well. Kawakami was accustomed to solving intellectual clashes by a pick-and-choose method and working for reconciliations and judicious blends of the values of competing philosophies. He was not unique in this regard. The Japanese eclectic approach to foreign culture, its tolerance for seemingly opposing religious sytems, is a distinctive feature of Japanese society.[12]

Logical consistency was not important; preservation of the unique essence of Japan was however of paramount importance. Kawakami's pragmatic approach to Marxism identified him as heir to a long history of Japanese cultural borrowing. He saw in Marxism a superior method of reform; he believed it was his privilege to select only that aspect of Marxism which Japan could fruitfully employ to reestablish the harmony and well-being of the social order. Fifteen hundred years earlier, the elite advisers of the emperor of Japan had in a similar fashion recommended importing Buddhism into Japan for its alleged magical properties. In both cases, the total thought system was eschewed in favor of certain selective elements deemed valuable to Japanese society. Kawakami's concrete, pragmatic and eclectic application of Marxism to modern Japan placed him squarely in the tradition of cultural assimilation that characterized his nation's history.

Marxism, however, was different from the other thought systems which the Japanese had adopted throughout their long history, because it was a total system of thought, embracing man, nature, and history in an intricately logical structure whose parts presumably could not be separated and rearranged, like a child's tinker toy. Rather, Marxism presented the Japanese intellectual with a new world view, rigorously materialistic, a monism that challenged the very core of Japanese society and social values with its own interpretation of the nature of social relations and historical development. Marxism insisted upon an end to the coexistence of ideas, in Maruyama Masao's expression, and hence required of the would-be Japanese Marxist a "bracing up" of his mental processes.[13] Because Marxism posed as a science, this demand was irresistible.

More than a methodology, Marxism in Japan became an orthodoxy with the authority of scriptures. *Capital* was the bible for Kawakami and numerous other Marxists as well. If Marxism became an intellectual tyranny, it is because it was called upon to serve several functions in Japan. Professor Maruyama believes that the "immense significance of Marxism in the history of ideas in Japan" lies in the fact that it incorporated at least three separate European thought systems: the conscience of Christianity, the logic of modern rationalism (in the form of the dialectic), and the empiricism of social science.[14] In the condensed period of time in which Marxism was allowed to flourish in Japan, it embodied both the fundamentalist fervor of Christianity and the methodology of social science, with the result that Marxism came to represent, in Robert Scalapino's words, a "combination of faith and science, or perhaps faith in 'science.' "[15] Both the intellectual discipline of the theorist and the fervent loyalty of the believer were involved in Kawakami's academic faith in Marxism.

The objectivity and discipline required of the modern scholar was sometimes more than Kawakami could bear. In his autobiography, he was scrupulous about citing documents or warning the reader that a certain date might be wrong because the document referred to was not immediately available, but at the same time he exercised little restraint in delivering up pieces of gossip or airing petty grievances. He complained, for example, that Suzuki Yasuzō used to dry his "smelly socks" on the communal brazier.[16]

Still it cannot be denied that Kawakami demonstrated throughout his career an openness to criticism and a willingness to correct his theoretical errors that freed him from the kind of dogmatism and name calling that often passed in his circles for academic debate. It was this intellectual integrity that led him to make numerous conversions throughout his life in the interests of correcting his scholarship but to resist merely opportunistic, ideological about-faces.

Other factors, too, account for Kawakami's refusal to renounce his academic faith. Unlike numerous other political prisoners who had to contend with the tearful implorings of family members, Kawakami gained encouragement for his defiance from those closest to him. His uncle urged him to accept his five-year prison term with fortitude, saying, "You should be able to keep up your spirits even if they give you ten years."[17] His brother, Sakyō, attended the court trial and brought him gifts when he was in jail. Although she disapproved of her husband's politicking, Kawakami's wife also stood loyally behind him in his resistance to the demands made upon him. She wrote him regularly, sent him food, met him at the gate when he was released from prison, and nursed him through his final years of life. When asked in her later years why her imprisoned husband had resisted police pressure to renounce Marxism, Mrs. Kawakami replied simply, "We are from Chōshū."[18]

An unbending will, stubborn defiance, and the capacity for independent action—these were the traits of character instilled in Kawakami by his Chōshū samurai upbringing. Kawakami was consciously aware of being a carrier of the shishi tradition: Yoshida Shōin, his childhood hero, had also gone to prison for refusing to yield to authority. As a young man, Kawakami had wept reading Yoshida's letters from prison to his mother, thirty years later, he was writing equally poignant letters and poems from prison to his own mother.[19] Kawakami's ascetic life style and his devotion to scholarship might have been patterned after Shōin's, whose idealism and sense of mission Kawakami shared. Someone who knew Kawakami personally in his later years remarked that he resembled Yoshida Shōin in the "strength of his will and the heat of his feelings."[20] Kawakami himself came to justify his revolutionary ideology in terms of the ethos of the shishi. "Even after I became a Marxist," he wrote,

> I never forgot Japan or disliked the Japanese. On the contrary, because I desired the welfare of the Japanese as a whole and the prosperity of Japan as a nation, I subscribed all the more ardently to the reconstruction of this country in a soviet system. Even if a man hopes for the defeat of his own country in war, that does not necessarily deny the fact that he is a patriot ... Those *shishi* who aspire to the revolutionary reform of the political system of the nation sometimes ardently wish for their country's defeat. Their attitude is derived from sincere feelings of patriotism in its true sense.[21]

Finally, even in his prison uniform, Kawakami's scholar status earned him the respect of his jailers. He was not subjected to physical violence, though prison fare was punishment enough. (He had to be hospitalized for a time, suffering from malnutrition and his chronic stomach ailments.) In other ways as well, the burden of imprisonment was lightened for the well-known professor. The warden invited a number of visitors to meet his famous prisoner and allowed him to work in the prison library, where he was urged to read in order to see the error of his Marxist ways. Kawakami was assigned the familiar task of translating foreign language works, and he received permission from prison authorities to keep a diary, entitled *Leisurely Meditation in Prison (Gokuchū zeigo)*.[22]

In the pages of his diary, Kawakami made a final effort to reconcile the underlying ethical source of his political behavior with the scientific truth of Marxism. What was the relation between the two truths—religion and science—he had pursued throughout his career? The religion he treated in his prison diary was not the historical or revealed religions of the Western world. Kawakami joined with Marx in labeling these organized religions the opiate of the masses and confined his own discussion mainly to Buddhism.

His basic argument, essentially unchanged since 1912, was that there were two realms of knowledge, different in the nature of the truths they contained and in their methods of uncovering these truths. The first concerned objective knowledge of the external world, for which the methods of science were appropriate. The second involved the subjective knowledge of the inner realm of self, for which the processes of intuition were required. In this second kind of knowledge—"consciousness of consciousness itself"—the individual attained unity with the absolute, an experience which Kawakami variously described as "hearing the Way" or "attaining Buddhahood by seeing one's own nature."

Despite this dualism which he established, Kawakami insisted on defending himself as a materialist. He argued that consciousness was a function of the brain, and that the brain was a physical entity. Consciousness depended ultimately upon the primary reality of material existence.[23]

This distinction between two realms of truth may be valid as far as it goes. But it does not go far enough. The frequently recalled religious experience of Kawakami's student days left him with more than a taste of mystical unity. It supplied him with an ethical basis for action in the world. Kawakami was avoiding the main issue, which is whether religious experiences or moral codes can exist independent of economic class determinants.

To the Marxist it is unimportant what the actor subjectively thinks he believes in. Benjamin Schwartz sums up the matter succinctly: "Man may indeed act out of a sense of duty, but his concept of where his duty lies is determined by the economic system in which he lives."[24] Therefore, it makes little sense to separate organized religion from religious truth, relegating the first to the category of ideology, and trying to rescue the second from the clutches of materialism. Morality, according to Marxism, cannot serve as a source of action as long as men and their ideas are determined by the mode of production of the society in which they live.

In the hours of reflection afforded by his prison term, Kawakami came to appreciate the ultimate incompatibility of his two world views, and he concluded that he was really a special (*tokushu*) Marxist.[25] The special quality of his Marxism locates him as a Japanese intellectual in a transitional age whose deepest spiritual needs Marxism ultimately failed to satisfy.

How should we comprehend the life of this "special Marxist"? In his memoirs, Kawakami identified as the consistent theme of his life his search for the Way. His search for the proper public cause to which he could completely dedicate himself led him through many roles: the religious prophet of the Selfless Love movement, the defender of the Japanese nation, the patron of Marxism and, finally, the champion of communism. In all four roles, he fought the common enemy—selfishness—most brazenly represented by the capitalist ethos.

Kawakami fought the enemy on the level of ideas and he fought it as a personal foe. When he preached selfless love, he gave all his clothes to the poor. For as long as he believed words could move the world, he gave himself over to scholarship. When he became convinced that the battle had to be won at the polls, he joined the proletarian party movement. When he realized the battle could only be waged underground, he worked for the Communist party.

The object of these battles was often blurred by the veil of Kawakami's own personal, religious search and struggle. The reader may wonder whether Kawakami was trying to save the working class from material impoverishment or himself from spiritual impurity, since he placed far greater importance on the purity of his will than on the outcome of his efforts. He asked to be judged on the basis of sincerity, not results. Thus he never simply made a decision; he "firmly resolved," or "girded his loins," or "dedicated himself wholeheartedly" to a particular course of action. He lived his life in the stirrup, poised to charge ahead into combat, as soon as he could figure out which side to be on.

Although Kawakami's contribution to Japanese Marxist scholarship was prodigious, he is remembered by friends today primarily as a poet and religious seeker (*kyūdōsha*).[26] The over one hundred members of the Tokyo and Kyoto Kawakami Hajime Commemorative Societies, who meet three times a year on the anniversaries of Kawakami's birthday, his death, and his release from prison, to reminisce about his personal qualities seem more determined to keep alive the memory of the man than the scholarship of the Marxist.[27]

At least one critic, however, has challenged Kawakami's description of his own life as a search for the Way. Ōkōchi Kazuo suggests that,

> Dr. Kawakami's successive "settlement of accounts" with his own academic theories ... "with the ease of a Nō player exchanging one mask for another," became for him the process of the "search for the Way." To Dr. Kawakami, such resolute transformations of self were proof of emergence and progress, but it is possible that what he called ... the fruits of this "search for the Way" were only dead bodies piled one on top of the other within the continuous person of the Doctor.[28]

It is indeed true that for all his preaching about altruism and self-sacrifice and the evils of individualism, Kawakami remained a highly individualistic and at times extremely selfish person. His insistence on a selfless life placed cruel burdens on family members who were all too often expected to sacrifice a great deal of their own time, energy, resources, and emotional needs in order to help Kawakami achieve his idealistic ends. And, too, for all his preaching about universal love, Kawakami often failed in his relationships with others. He was by his own admission a poor judge of character. His high hopes were repeatedly dashed against the realities of human frailty. Friends disappointed him, allies betrayed him, students turned against him.

His expectations were often unrealistic and naive, or influenced by his own personal desires. When hurt he sought revenge, usually with the power of his pen. After his political ally, Ōyama Ikuo, broke away from the Communist movement, he became, in Kawakami's eyes, bourgeois. In his autobiography, Kawakami later described the proletarian party leader living in a middle-class, Western style of life, and to illustrate his decadence, depicted him seated at his dining room table, complaining about his poverty, while indulging his guests in chunks of butter without bread.[29]

For years Kawakami wrote about the problem of poverty without ever having known any of the downtrodden or lived among them. This is not to suggest that he was insincere; rather, he was trapped throughout most of life in mental abstractions. In his own respect for learning, in his social responsibility "at a distance," and in his confidence in the healing powers of virtue, he stood at the end of a long line of equally respectable, Confucian-type gentleman scholars. That he could break away from that mold, however awkwardly, to merge with the masses when the time came is a tribute to his sincerity, just as his political failure is a measure of how much separated he was from them.

As a scholar, Kawakami Hajime was a courageous activist, and as an activist, he was a cautious scholar. Always in danger of falling between two worlds, he was a constant embarrassment to both. When he published encomiums on such inflammatory subjects as revolution, he strained the patience of the Ministry of Education, and when he led discussions of Marx's *Capital* at study groups on the campus of Kyoto Imperial University, administrative officials shifted uneasily in their seats. At the same time, when he tried to lecture on *Capital* at labor union rallies—his way of being *engagé*—he lost his audience within the first five minutes,[30] and he failed in his one attempt to campaign for a Diet seat on the Labor–Farmer ticket.

Refusing to cater to his bookish disposition, he forced himself into revolutionary activity, but his real talent lay in journalism, where he experienced his greatest popular success. Here he could reign as the conscience of a generation over a large and literate public, reaching them through words far better than he could through actual deeds. He was at his best as a moralist, a public guardian of goodness, alternating between sentimentality and outrage, crusading for the poor with pen and paper as his weapons. He felt most alive when he was morally indignant; and like his childhood heroes, Yoshida Shōin and Tanaka Shōzō, he thrived on emotional appeals for dedicated self-sacrifice in the interests of the collective good.

Kawakami was as aware of his following as any political leader would be. When he changed his mind about a particular subject—and this was often—he provided public notice of his *volte face*. Confessions, resolutions, personal decisions were shared with his audience in a manner so frank as to suggest that Kawakami considered his life a piece of public property. The published autobiography offended some with its candor and prompted others to compare it with Rousseau's *Confessions*.[31] What is important to remember is Kawakami's need to justify his life to himself and to the Japanese people.

Prevented by the discipline of scientific socialism from allowing his subjectivism to spill over into the pages of economic analysis, Kawakami took to using the introductions to his books to explain to his readers the course of his travels to Marxism. The autobiography was his final attempt to plead his case before his invisible judges; to counter charges that he was weak-willed or inconstant; to dispel the impression that his numerous intellectual changes were guided by opportunism; to demonstrate that he did not hold a blind, uncritical faith in Marxism; and finally, to prove to his public, and perhaps to himself, the righteousness of his actions and the unity of his life.[32]

Kawakami Hajime's autobiography tells the story of a member of the traditional Japanese elite who, caught in the unsettling changes of the past one hundred years, persistently tried to adapt old values to the new circumstances of his existence. If Yoshida Shōin demonstrated how to rework feudal values into nationalist sentiment, Kawakami Hajime tried to broaden the scope of the shishi ethos even further so that it embraced all of mankind.

Modernizing Japanese ethics was only one aspect of the task Kawakami confronted as an ethical economist. His ethics of selflessness also represented an attempt to temper Western values of individualism with Japanese altruistic morality. His duty-bound individualism reflected his desire to create a new ethical ideal that preserved the self-denying aspects of traditional Japanese morality while recognizing the needs of the modern individual.

Simple dichotomies between Japan and the West of the sort Kawakami frequently drew only partially describe his plight as a modern Japanese philosopher and social scientist. Contradictory elements abounded within Western culture itself; and Kawakami Hajime's life work involved sorting and selecting from among the complex heritage of European and American societies. The contradiction between the ethics of Christianity and the ethos of capitalism was as stark as that between Japanese morality and Western individualism. Moreover, the conflict within Kawakami's own personality between selflessness and selfishness, or between self-denial and individual self-assertion cautions us against accepting any easy distinctions between traditional and modern, or Japanese and Western in explaining Kawakami's philosophical predicament.

Kawakami's Marxist critics overlooked this complexity. His vocabulary of Confucian and Buddhist words made him sound old-fashioned in the eyes of his intellectual peers and students, who did not understand or approve of the way he tried to use "tradition to overcome traditionalism."[33] Responding to their criticism, he struggled to translate his values into a more modern idiom. To the extent that Marxism shared certain similarities with his own ethical views, he succeeded: his ethics of absolute selflessness, dressed as revolutionary zeal, passed for a while as Marxist orthodoxy, and his scholarly lectures on *Capital* temporarily served the requirements of political practice.

But, more than a Marxist, Kawakami was a modern-day samurai in search of a principle to which he could dedicate his life. His loyalty to a cause, or to the truth, like loyalty to a lord, had to be unconditional, absolute, thorough. Anything less would have been selfish. And un-Japanese. *Isn't Revolution Fun?* is the title of Fukumoto Kazuo's autobiography. Kawakami became a revolutionary because it was not fun. It was a duty. It was painful, and it demanded the utmost in sacrifice of him. Out of these Japanese values, Kawakami fashioned his revolutionary ideology; his ethics of selflessness imparted meaning to his Marxist activities and prison years, enabling him to believe that he had fought as heroically as the samurai in battle and that he had not lived and would not die in vain.

The imagery of the battlefield appears in one of Kawakami's last written works, a poem entitled "On the Death Bed," published in *Red Flag* shortly before his death. The poem was dedicated to two imprisoned Communist party comrades, Shiga Yoshio and Tokuda Kyūichi, but Kawakami was perhaps thinking of himself as well, when he wrote: "Fighting, fighting, you survived,/And in the end, your spirit was not broken."[34]

notes/bibliography glossary/index

notes

1. A Young Man of Meiji Japan

1. For a history of this period, see William Beasley, *The Meiji Restoration* (Stanford, 1972). For a study of Chōshū's role in Bakumatsu politics, see Albert M. Craig, *Chōshū in the Meiji Restoration* (Cambridge, 1961).
2. Craig, *Chōshū in the Meiji Restoration*, pp. 321–322.
3. Kawakami Tetsutarō, *Nihon no autosaidā* (Tokyo, 1961), p. 116.
4. Kawakami Hajime, *Jijoden, Kawakami Hajime chosakushū* (Tokyo, 1964), VI, 10. Hereafter cited as *Jijoden*.
5. One koku was equivalent to 4.96 bushels of rice.
6. Beasley, *The Meiji Restoration*, p. 29. At the time of the Restoration, there were approximately 925 Chōshū samurai families with incomes of less than 40 koku out of a total of about 2,600. Although the designation of upper, middle, and lower samurai status varies according to the criteria used, Beasley believes 20 koku was the dividing line in Chōshū between middle-ranking (*hirazamurai*) and lower samurai. The Kawakami family's residence outside the castle town suggests they were equivalent to *gōshi* (rural samurai), one of three subdivisions within the lower samurai class.
7. "Omoide," *Chosakushū*, IX, 201. Hereafter cited as "Omoide."
8. Ōtsuka Yūshō, *Mikan no tabiji* (Tokyo, 1961), IV, 150.
9. *Jijoden*, p. 10. Wherever possible I have used the Western way of calculating age. By Japanese calculations, Kawakami's mother was seventeen when she married, but according to Kawakami, "to be precise, she was fifteen years and seven months old." See "Omoide," p. 202.
10. *Jijoden*, p. 10.
11. Ibid. Sunao's second wife was Inoue Shie.
12. *Jijoden*, pp. 10-11.
13. Sumiya Etsuji, *Kawakami Hajime* (Tokyo, 1962), p. 13.
14. *Jijoden*, pp. 12–17.
15. Ibid.
16. V. S. Pritchett, "The Strength of an Injured Spirit," *New York Review of Books,* November 2, 1972, p. 8.
17. *Jijoden*, p. 24.
18. Ivan P. Hall, *Mori Arinori* (Cambridge, 1973), pp. 26 and 28; Tokutomi Iichirō, *Kōshaku Yamagata Aritomo den* (Tokyo, 1969), p. 54.
19. *Jijoden*, p. 24.
20. Kawakami Hajimes, "Sobo no tsuioku," *Kawakami Hajime chosakushū*, VIII, p. 476.
21. *Jijoden*, p. 23.
22. Ibid. Loss of the pleasure of breast feeding is considered, by Erik Erikson, one of the early childhood traumata universal to man. Erikson describes the reaction of rage common to Sioux Indian children when they are weaned. See his *Childhood and Society* (New York, 1963), pp. 50–51. Studies of

Japanese child rearing report a common pattern of temper tantrums among children as old as eight. Ruth Benedict considered temper tantrums to be acts of aggression against mothers and grandmothers and viewed them as normal until the age of six. See her *Chrysanthemum and the Sword* (New York, 1972), p. 264.
23. *Jijoden*, pp. 24–27.
24. Ibid., p. 29. The fact that Kawakami always had his own way as a child developed in him the "determination to make his will prevail whatever the circumstances," according to Kazuko Tsurumi, *Social Change and the Individual* (Princeton, 1970), p. 65, quoting Mimpei Sugiura, "Kawakami Hajime ron," *Shisō no kagaku* 44:25 (1965). Japanese mothers today, however, seem to operate on the opposite assumption, namely, that letting the child have his own way, instead of thwarting him, will minimize his assertiveness. See Ezra Vogel, *Japan's New Middle Class* (Berkeley, 1963), p. 245.
25. Ōtsuka Yūshō, *Mikan no tabiji*, IV, 155.
26. Hamura Shizuko, "Chichi no koto," Horie Muraichi, ed., *Kaisō no Kawakami Hajime* (Tokyo, 1948), p. 166.
27. Kawakami Tetsutarō, *Nihon no autosaidā*, p. 116.
28. *Jijoden*, p. 6.
29. Ibid., p. 9.
30. Ibid., p. 8.
31. Ōtsuka Yūshō, *Mikan no tabiji*, IV, 152–153.
32. Hamura Shizuko, "Chichi no koto," p. 166.
33. Furuta Hikaru, *Kawakami Hajime* (Tokyo, 1959), p. 27. "Bōchō" was written with the characters for Suo and Nagato, two former Mōri domains active at the time of the Restoration. These were later combined to form Yamaguchi Prefecture.
34. *Iwakuni shishi* (Iwakuni, 1957), pp. 1108–1110; "Jijoden," pp. 34–36; *Yamaguchi-ken bunka-shi* (Yamaguchi, 1959), pp. 109–110.
35. *Jijoden*, pp. 36–37.
36. Furuta, *Kawakami Hajime*, pp. 26–27. According to Kawakami, the four other schools were in Yamaguchi City, Hagi, Toyoura, and Tokuyama. An official history of education in Yamaguchi Prefecture lists only four, omitting Tokuyama. See *Yamaguchi-ken bunka-shi*, pp. 109–110.
37. Sumiya Etsuji, *Kawakami Hajime*, p. 22.
38. *Jijoden*, pp. 38 and 44.
39. Ibid., pp. 37–39.
40. Ibid., p. 43.
41. Ibid., pp. 45–47.
42. Ibid., pp. 49–54.
43. Ibid., p. 48.
44. Ibid., pp. 54–55. Fūgetsu was also the name of a famous Tokyo bakery. When Kamakami discovered the coincidence, he abandoned the pseudonym.
45. Ibid., pp. 46–48; 54–55.
46. Ibid., pp. 27–30; 55.
47. Sakuda Shōichi, *Jidai no hito: Kawakami Hajime* (Osaka, 1949), p. 6.
48. *Jijoden*, pp. 55–56.
49. Craig, *Chōshū in the Meiji Restoration*, pp. 156–164.
50. Masaaki Kōsaka, *Japanese Thought in the Meiji Period*, tr. David Abosch (Tokyo, 1958), p. 205.

51. *Jijoden*, p. 56.
52. Ibid., pp. 55–57.

2. Crisis

1. The population of Tokyo in 1900 was approximately one and a half million. The city had a circumference of thirty miles and a total area, including suburbs, of one hundred square miles. The population of Japan was about forty-three million. See Shibusawa Keizō, *Japanese Life and Culture in the Meiji Period* (Tokyo, 1959), p. 332; Frank Brinkley, ed., *Japan Described and Illustrated by the Japanese* (Boston, 1897), I:9n; Yazaki Takeo, *Social Change and the City in Japan* (Tokyo, 1968), p. 420.
2. *New York Times*, November 24, 1895, 29:3; Shibusawa Keizō, *Japanese Life and Culture in the Meiji Period*, pp. 145, 202–203; F. H. Balfour, "Court and Society in Tōkyō," *Transactions and Proceedings of the Japan Society, London* 3:54–57 (1893–94); Yazaki Takeo, *Social Change and the City in Japan*, pp. 359–360.
3. Shibusawa Keizō, *Japanese Life and Culture in the Meiji Period*, pp. 25–27; Balfour, "Court and Society in Tōkyō," pp. 65–66. Balfour lamented the Japanese noblemen's custom of wearing kimono, *haori*, and *geta* with English bowler hats.
4. *New York Times*, May 31, 1897, 3:2.
5. *Jijoden*, p. 45.
6. "Omoide," p. 260.
7. Ibid., p. 252.
8. Ibid., pp. 280–285.
9. Kawakami Hajime, "Kakuhitsu no ji," *Shakaishugi hyōron* (Tokyo, 1906), 1st ed., reprinted in Ōuchi Hyōe, ed., *Kawakami Hajime* (Tokyo, 1964), p. 137.
10. Fukui Kōji, "Ano koto, kono koto," Horie Muraichi, ed., *Kaisō no Kawakami Hajime*, pp. 228–229.
11. "Omoide," pp. 316–318.
12. Ibid., p. 409.
13. "Kakuhitsu no ji," p. 132. Six ideographs were omitted from the original text just after Kawakami wrote, "I loved seven women, but the last love was a certain woman with whom one night . . ."
14. Ibid., p. 137.
15. "Omoide," pp. 216–217.
16. Henry D. Smith, II, *Japan's First Student Radicals* (Cambridge, 1972), p. 7.
17. Ōuchi Hyōe, *Keizaigaku gojūnen* (Tokyo, 1959), p. 1. Ōuchi estimates that in 1908 there were about four times as many candidates in law and government as in economics.
18. Ibid.
19. Ibid., pp. 9–10.
20. Ibid., pp. 30–32. Matsuzaki was later diagnosed as suffering from dementia praecox, a form of insanity characterized by incoherent thought and action.
21. *Shakaishugi hyōron*, p. 19.
22. "Omoide," p. 317.
23. *Jijoden*, p. 348.
24. Maruyama Masao, "Patterns of Individuation and the Case of Japan: A Conceptual Scheme," Marius Jansen, ed., *Changing Japanese Attitudes toward Modernization* (Princeton, 1965), p. 509.

25. Kōsaka Masaaki, *Japanese Thought in the Meiji Period*, p. 301.
26. "Omoide," pp. 219–220.
27. Ibid., p. 219, from Matthew 5:39–42.
28. Kazuko Tsurumi, *Social Change and the Individual Japan* (Princeton, 1970), p. 70.
29. "Omoide," p. 219.
30. Kishimoto Hideo, *Japanese Religion in the Meiji Era*, tr. John F. Howes (Tokyo, 1956), p. 286.
31. "Omoide," pp. 260–261.
32. Ibid., pp. 220–221.
33. Ibid., pp. 257–259.
34. "Kakuhitsu no ji," p. 132.
35. Nobody in Kawakami's immediate family had ever been to Tokyo and so they had never been exposed to the social criticism that had awakened Hajime's conscience. His new world was very different from their own provincial environment. When his father finally visited the capital in 1905 to celebrate his retirement from office, his enthusiasm for modern life was expressed in his pleasure over the roof garden of a downtown department store. This childlike quality also appeared in his desire to see a grossly overweight child whom he had read about in a Tokyo newspaper. He actually traveled to the child's neighborhood and stood outside the house waiting, until the child's mother demanded to know what he wanted, whereupon he quickly left. Hajime's high expectations of men in the public trust may have reflected the imprint of his father's own exemplary behavior while in office, but Sunao, whose tranquility was based in part on his naivete about the ways of the world, probably could not genuinely comprehend Hajime's distraught mind, though his gentle, tolerant nature must have been some comfort to Hajime. See Otsuka Yūshō, *Mikan no tabiji*, IV, 153-155.
36. "Omoide," pp. 218–219.

3. Meiji Dropout

1. Kawakami Hide, *Rusu nikki* (Tokyo, 1967), pp. 369–370.
2. Ibid., p. 365.
3. Tsuda Seifū, "Kawakami Hakase no ningensei," in Suekawa Hiroshi, ed., *Kawakami Hajime kenkyū* (Tokyo, 1965), p. 290.
4. *Jijoden*, pp. 347–348.
5. Letter to Katayama Sen, printed in *Rōdō sekai*, 6.15 (September 13, 1902).
6. "Omoide," pp. 284 and 264–265; "Kakuhitsu no ji," p. 133. He was also appointed to a professorship at the Peers School and a lectureship at Senshū School. See Sumiya Etsuji, *Kawakami Hajime*, pp. 321–322.
7. Ōyama Ikuo, *Keizaigaku gojūnen*, p. 18. For a fuller treatment of the German school of national economics, see Chapter 4.
8. Kawakami Hajime, *Keizaigaku genron*, quoted by Ōkōchi Kazuo, "Kawakami Hajime to kyūdō no kagaku," *Chūō kōron*, 3:435 (March, 1965).
9. Ibid.
10. Ibid.
11. *Shakaishugi hyōron*, 6th ed. (Tokyo, 1905), p. 19.
12. Ibid., p. 39.
13. *Shakaishugi hyōron*, appendix, p. 176. Entire passage italicized in the original.

14. Edwin R. A. Seligman, *The Economic Interpretation of History* (New York, 1907), pp. 1–3. Originally published in *Political Science Quarterly* 16 (1901) and 17 (1902).
15. Ibid., p. 38.
16. *Jijoden*, pp. 113–114. We do not know why he chose this particular book to translate at this time. Its newness and the author's suggestion that it presented the most up-to-date theory of modern science must have appealed to Kawakami, who was himself seeking to discover the great principles of economics.
17. *Shakaishugi hyōron*, pp. 178–180.
18. Robert Jay Lifton, "The Struggle for Cultural Rebirth," *Harpers Magazine* (April 1973), p. 90.
19. *Shakaishugi hyōron*, p. 115.
20. Ibid., p. 178.
21. Ibid., pp. 181–182.
22. Ibid., pp. 182–183.
23. "Omoide," pp. 193–194.
24. Ibid., pp. 221–222, and "Kakuhitsu no ji," p. 134. Kawakami was also troubled by the love affair that developed between the two cousins while they were staying at his house.
25. *Jijoden*, pp. 113–114.
26. Ōuchi Hyōe, quoted by Furuta, *Kawakami Hajime,* p. 81. Ōuchi decided to study economics after reading this book. See Ōuchi Hyōe, *Keizaigaku gojūnen*, p. 8.
27. Amano Keitarō, *Kawakami Hajime hakase bunkenshi* (Tokyo, 1956), p. 11.
28. Kawakami Hajime, *Keizai gakujō no kompon kannen* (Tokyo, 1905), pp. 17–19.
29. "Omoide," p. 223.
30. *Shakaishugi hyōron*, pp. 118–124.
31. "Omoide," p. 223.
32. "Kakuhitsu no ji," p. 135.
33. *Shakaishugi hyōron*, preface, pp. 1-3. Pretending to be suffering from a serious illness, he had notified friends that he was returning home to Yamaguchi and then he had remained in Tokyo to write the articles under an assumed name. As a final piece of deception, he informed his readers in the very first article that he had recently returned from a trip abroad.
34. Ōuchi Hyōe, "Kaisetsu," in Ōuchi Hyōe, ed., *Kawakami Hajime*, pp. 17–18. See also Ōuchi's *Keizaigaku gojūnen*, pp. 2–3. Ōuchi, who was a middle school student when Kawakami's *Critique of Socialism* was published, recalled, "When I read it, I learned for the first time that there was something called society and that there are scholarly debates about such things."
35. *Shakaishugi hyōron*, pp. 65–66.
36. Ibid., p. 43.
37. Ibid. The final paragraphs of the last installment in Kawakami's series of articles were omitted from the sixth edition of the book. For his criticism of the Christian socialists, see the version reprinted in Ōuchi Hyōe, ed., *Kawakami Hajime*, p. 128.
38. "Omoide," p. 223.
39. Ibid.; "Kakuhitsu no ji," p. 136.
40. Leo Tolstoy, *My Religion,* tr. Leo Wiener. *The Complete Works of Count Tolstoy* (London, 1904), XVI, 10.

41. "Omoide," pp. 223–224.
42. Yoshida Kyūichi, "Muga ai undō to Kawakami Hajime," *Nihon rekishi,* 165:4–5 (March 1962).
43. "Omoide," p. 229.
44. "Kakuhitsu no ji," pp. 129–138.
45. Kawakami Hajime, "Muga ai no shinri," *Kawakami Hajime,* Gendai nihon shisō taikei, XIX, 138.
46. Ibid., p. 145.
47. "Omoide," pp. 234–235. Scattered references to Kawakami's selfless love period may also be found in his "Jijoden," pp. 59–84 and "Omoide," pp. 215–250.
48. For an attempt to describe the experience of enlightenment analytically, see Richard DeMartino, "The Human Situation and Zen Buddhism," in D. T. Suzuki, Erich Fromm, and Richard DeMartino, *Zen Buddhism and Psychoanalysis* (New York, 1960), pp. 142–171.
49. Sumiya, *Kawakami Hajime,* pp. 98–99.
50. Erik Erikson, *Young Man Luther* (New York, 1962), pp. 37–39.
51. "Muga ai no shinri," pp. 142–143.
52. William T. DeBary, *Self and Society in Ming Thought* (New York, 1970), pp. 12–16. DeBary points out that even though such phenomena, which were widespread, bore similarities with Zen experience, they may nonetheless "fall into a broader category of mysticism which need not always be labeled 'Ch'an.' Indeed the possibility of a distinctive Confucian mysticism can by no means be ruled out."
53. "Omoide," pp. 244–245.
54. *Jijoden,* p. 76.
55. The exact date of Kawakami's entrance into the Garden of Selflessness is difficult to determine. Furuta Hikaru gives the date as December 8, 1905, but this is too early, since Kawakami's "great death" occurred between December 8 and 9, and afterward, he briefly returned home to Yamaguchi. Kawakami writes in his memoirs that he entered the sect "at the end of year" and that the house was completed on January 4, 1906. See "Omoide," pp. 245–246.
56. Kawakami Hajime, "Jinsei no kisū," quoted in Sumiya Etsuji, *Kawakami Hajime,* p. 104.
57. "Had I known clearly from the start what Itō's emphasis was, I would never have leaped into the Garden of Selflessness," Kawakami wrote in his autobiography. See *Jijoden,* p. 66.

4. The "Way" in the Modern World

1. Yoshino's review appeared in *Kokka gakkai zasshi* 19.8 (August 1905).
2. Sumiya Etsuji, *Kawakami Hajime,* p. 102.
3. Ōuchi Hyōe, *Keizaigaku gojūnen,* p. 41.
4. Furuta Hikaru, *Kawakami Hajime,* p. 85n.
5. Sumiya Etsuji, *Nihon keizaigaku shi* (Kyoto, 1958), pp. 8ff.
6. Ōuchi Hyōe, *Keizaigaku gojūnen,* p. 17. On the introduction of the German school into Japan, see also Sumiya Etsuji, *Nihon keizaigaku shi,* pp. 151–161.
7. For a clear and interesting account of the German school and its relation to other schools of economics, see Wesley C. Mitchell, *Lecture Notes on Types of Economic Theory* (New York, 1949). Much of the discussion which

follows is based on this source. For a more critical treatment of the German school, see Joseph A. Schumpeter, *History of Economic Analysis* (New York, 1954).
8. Schumpeter, *History of Economic Analysis*, p. 803n.
9. Ibid., p. 802.
10. On the meaning of science and research, see Schumpeter, p. 807.
11. Sumiya Etsuji, *Nihon keizaigaku shi*, p. 250. For a fuller treatment of the association, see his *Nihon keizaigaku shi no hitokoma* (Tokyo, 1948), pp. 95–115, and Ōuchi Hyōe, *Keizaigaku gojūnen*, pp. 43–92.
12. Sumiya Etsuji, *Nihon keizaigaku shi*, p. 171. For an English translation of the prospectus, see Hyman Kublin, *Asian Revolutionary, The Life of Katayama Sen* (Princeton, 1964), p. 137.
13. Ōuchi Hyōe, *Keizaigaku gojūnen*, p. 51. See also Sumiya Etsuji, *Nihon keizaigaku shi*, p. 253.
14. Sumiya Etsuji, *Nihon keizaigaku shi*, p. 178, quoting Ōuchi Hyōe.
15. For their "Letter of Explanation" ("Benmeisho"), see Sumiya Etsuji, *Nihon keizaigaku shi*, pp. 171–173.
16. Kawakami Hajime, *Shakaishugi hyōron*, pp. 65–66. Reprinted in *Kawakami Hajime chosakushū*, I.
17. *Jisei no hen*, p. 44.
18. Kawakami Hajime, "Keizai to dōtoku," from *Nihon keizai shinshi*, reprinted in *Kawakami Hajime chosakushū*, I, 457–458.
19. Ibid.
20. Kawakami Hajime, *Nihon nōseigaku* (Tokyo, 1906), p. 380. The absolute number of workers in agriculture actually remained constant until 1930. See Kazushi Ohkawa and Henry Rosovsky, "A Century of Japanese Economic Growth," in William Lockwood, ed., *The State and Economic Enterprise in Japan* (Princeton, 1965), p. 73.
21. Ohkawa and Rosovsky, p. 75.
22. Ronald P. Dore, *Land Reform in Japan* (London, 1959), p. 61.
23. Kawakami Hajime, *Nihon sonnō ron* (Tokyo, 1905), p. 75. Reprinted in *Chosakushū*, I.
24. *Nihon nōseigaku*, pp. 125–126 and passim.
25. *Jisei no hen*, pp. 102–103.
26. *Nihon nōseigaku*, pt. 2.
27. *Sonnō ron*, quoted by Uchida Yoshihiko, "Meiji makki no Kawakami Hajime," in Yamada Moritarō, ed., *Nihon shihonshugi no sho mondai* (Tokyo, 1960), p. 182.
28. Ibid., pp. 183–184.
29. *Sonnō ron*, p. 148.
30. Ibid.
31. *Nihon nōseigaku*, p. 159. While it was true that the rural birth rate was higher than the urban birth rate, it was not true that the urban mortality rate exceeded the urban birth rate. See Frederick W. Poos, "Kawakami Hajime, 1879–1946: An Intellectual Biography," Ph.D. diss. Stanford University, 1957, p. 33.
32. Ibid., from *Nihon nōseigaku*, pp. 148–150; 388–399.
33. Ibid., from *Nihon nōseigaku*, pp. 158–165.
34. "Keizai to dōtoku," *Nihon keizai shinshi* 1.2:26 (April 18, 1907), reprinted in *Chosakushū* I, 457.

35. Uchida Yoshihiko develops this theme of the conflict between Kawakami as a moralist and as an economist in his "Meiji makki no Kawakami Hajime."
36. Robert Bellah, *Tokugawa Religion* (Glencoe, 1957), p. 5.
37. Quoted in Robert L. Heilbroner, *The Worldly Philosophers* (New York, 1953), pp. 46–47.
38. Delmar H. Brown, *Nationalism in Japan* (Berkeley, 1955), p. 152.
39. Kawakami Hajime, "Shūgi washo ni miraretaru Kumazawa Banzan no keizai gakusetsu," *Kokka gakkai zasshi* 21.10:1256 (October 1907). For a discussion of the Meiji revival of interest in Tokugawa thought, see Honjō Eijirō, *Edo-Meiji jidai no keizai gakusha* (Tokyo, 1962), pp. 189–206.
40. Ibid., p. 1255.
41. "Keizai to dōtoku," *Nihon keizai shinshi* 1.2:18 (May 3, 1907).
42. Kawakami Hajime, *Bimbō monogatari* (Iwanami, 1961), p. 154.
43. Quoted in Robert N. Bellah, *Tokugawa Religion*, p. 109
44. Kawakami Hajime, *Keizaigaku kompon kannen*, p. 30.
45. Sakuda Shōichi, *Jidai no hito*, pp. 30–31.
46. *The Analects of Confucius*, tr. Arthur Waley (New York, 1938), p. 88.
47. *Jijoden*, p. 122.
48. Ibid., p. 108.
49. Ibid., pp. 108–109.
50. *Keizai to jinsei, in Chosakushū*, VIII, 199.
51. *Bimbō monogatari*, p. 190.
52. Kawakami Hajime, "Bakumatsu jidai no shakai shugisha Satō Nobuhiro," *Kyōto hōgakkai zasshi* 4.10 (October 1909).
53. See, for example, Kawakami's article on Darwinism and Marxism in *Chūō kōron* 27.4:18 (April 1, 1912).
54. In 1906 Kawakami wrote an article under a nom de plume for the *Yomiuri* newspaper praising Kita Ikki for challenging "today's so-called scholar class." Kita had just published his *Kokutairon and Pure Socialism*, in which he attacked the theory of *kokutai*. See George M. Wilson, *Radical Nationalist in Japan: Kita Ikki, 1883–1937* (Cambridge, 1969), p. 41. In 1916, in his *Tale of Poverty*, Kawakami confessed, "I am afraid to be mistaken for or confused with the anarchists. Therefore, I avoid using the word socialism and use nationalism (*kokkashugi*) instead." (Iwanami edition, p. 109).
55. See, for example, "Shūgi Washo ni miraretaru Kumazawa Banzan no keizai gakusetsu," p. 1257.
56. *Jisei no hen*, p. 116. Reprinted in *Chosakushū*, VIII, 260.

5. Japan and the West

1. Sumiya Etsuji, *Nihon keizai gakushi*, p. 176.
2. *Sokoku o kaerimite* (Tokyo, 1915), p. 31.
3. Robert Bellah, "Japan's Cultural Identity," *Journal of Asian Studies* 4:588 (August 1965).
4. *Keizai to jinsei* (Tokyo, 1911), p. 309, reprinted in *Chosakushū*, VIII.
5. Ibid., p. 329.
6. Ibid., p. 189.
7. Ibid., p. 193.
8. Ibid., p. 194.
9. Ibid., pp. 204–205.
10. *Keizai to jinsei, Chosakushū*, VIII, 183.

11. *Jisei no hen,* quoted by Uchida Yoshihiko, "Meiji makki no Kawakami Hajime," in Yamada Moritarō, ed., *Nihon shihonshugi no sho mondai* (Tokyo, 1960), p. 194.
12. *Jisei no hen* (1911 ed.), p. 138. Entire passage italicized in original.
13. Ibid., quoted by Uchida, p. 194.
14. *Jisei no hen, Chosakushū,* VIII, 203.
15. "Keizai to jinsei," *Chosakushū,* VIII, 203.
16. *Jisei no hen* (1911 ed.), p. 163.
17. February 4, 1918, letter to Kushida Tamizō, in Ōuchi Hyōe, ed., *Kawakami Hajime yori Kushida Tamizō e no tegami* (Tokyo, 1947), p. 62. Hereafter cited as *Tegami*.
18. "Yuibutsu yori yuishinkan e," *Kokumin keizaigaku zasshi* (July-August 1912), quoted by Furuta Hikaru, *Kawakami Hajime,* pp. 98–99. These essays were included in Kawakami's *Keizaigaku kenkyū* (1912), an abridged version of which is reprinted in *Chosakushū,* I, 321–454.
19. April 6, 1913, letter to Kushida Tamizō, *Tegami,* pp. 31–32.
20. Shimazaki Tōson, "Etoranzee sho," in Suekawa Hiroshi, ed., *Kawakami Hajime kenkyū* (Tokyo, 1965), p. 283.
21. Ibid., p. 278.
22. *Sokoku o kaerimite,* pp. 239–246; Shimazaki Tōson, "Etoranzee sho," p. 275; Takeda Akira, "Oshū ryūgaka jidai no Kawakami san," in Horie Muraichi, ed., *Kaisō no Kawakami Hajime,* p. 8.
23. Takeda Akira, "Oshū ryūgaka jidai no Kawakami san," pp. 8–10; *Sokoku o kaerimite,* pp. 258–259.
24. Shimazaki Tōson, "Etoranzee sho," pp. 276–283.
25. *Sokoku o kaerimite,* p. 173.
26. Ibid., pp. 198–210.
27. Ibid., pp. 296–303.
28. Ibid., p. 311.
29. Ibid., pp. 31–32.
30. Ibid., pp. 198–210.
31. Ibid., pp. 29–31.
32. Ibid., pp. 94–95.
33. Furuta, *Kawakami Hajime,* p. 103.
34. "Omoide," p. 321.
35. *Sokoku o kaerimite,* p. 304.
36. "Kokondō zuihitsu," *Chosakushū,* IX, 409–410; "Omoide," p. 320.
37. "Omoide," pp. 319–320; Ōtsuka Yūshō, *Mikan no tabiji,* IV, 202–205.
38. *Sokoku o kaerimite,* p. 296.
39. Ibid., pp. 304–309.

6. The Road to Marxism

1. Kawakami Hajime, *Bimbō monogatari* (Iwanami, 1961), p. 9. Reprinted in *Kawakami Hajime chosakushū,* vol. II.
2. Ōuchi Hyōe, *Keizaigaku gojūnen* (Tokyo, 1959), p. 141.
3. *Bimbō monogatari* (Kōbundō, 1917), p. 127.
4. Ibid., p. 39.
5. *Bimbō monogatari* (Iwanami), p. 3.
6. Ibid., p. 39.

7. Ibid., pp. 77–78.
8. Ibid., p. 108.
9. Ibid., pp. 106–107.
10. Ibid., pp. 161–162.
11. Ibid., pp. 80–83.
12. Ibid., pp. 142–150, cited in Frederick Poos, "Kawakami Hajime (1879–1946): An Intellectual Biography," Ph.D. diss. Stanford University, 1957, p. 59.
13. Bimbō monogatari, p. 155.
14. Ibid., p. 90.
15. Ibid., pp. 44–60. Kawakami repeated James Haldane Smith's observations in Economic Moralism (1916) that European nations were moving toward greater government control of industry.
16. Ibid., pp. 109–110.
17. Ibid., p. 114.
18. Kawakami Hajime, "Sumāto no 'Ichi keizai gakusha no daini shisō,' " Keizai ronsō 4.2 (February 1917), reprinted in Chosakushū, X, 64.
19. Ōuchi Hyōe, "Keizaigakusha to shite no Kawakami Hajime," in Suekawa Hiroshi, ed., Kawakami Hajime kenkyū, p. 145.
20. Ibid.
21. Nihon no hyakunen (Tokyo, 1962), VI, 36.
22. Quoted by Sumiya Etsuji, Kawakami Hajime, p. 155.
23. Quoted in Lawrence Olson, Dimensions of Japan (New York, 1963), pp. 207–208. The estimate of the number of subscribers to the Asahi comes from Ōuchi Hyōe's "Kaidai," comments at the end of the 1961 Iwanami edition of Bimbō monogatari, p. 182. Many more actually read it in libraries and friends' houses. Members of the New Men's Society (Shinjinkai) at Tokyo Imperial University attributed their awareness of social problems to the influence of Bimbō monogatari. Henry D. Smith's study of about three hundred members shows that reading Bimbō monogatari was often a "crucial turning point" for young students, leading them away from brooding introspection and toward the study of social problems. See Smith's Japan's First Radical Students (Cambridge, 1972), p. 240.
24. Quoted by Furuta Hikaru, Kawakami Hajime, p. 117.
25. In his Second Tale of Poverty (Daini bimbō monogatari), originally published between 1929 and 1930, Kawakami repudiated the ethical approach to economic problems characteristic of the first Tale. See Chosakushū, II, 130–131, and Chapter 9.
26. Ōuchi Hyōe, "Kushida Tamizō, Marukusugaku no kakuritsusha," in Nihon no shisōka (Tokyo, 1968), III, 150–167.
27. Kushida had studied German at the Tokyo Foreign Language School before entering Kyoto Imperial University.
28. Ōuchi Hyōe, ed., Kawakami Hajime yori Kushida Tamizō e no tegami (Tokyo, 1947), pp. 41–47.
29. Kushida Tamizō, "Kawakami kyōju no 'Shahi to hinkon' wo yomite," Kokka gakkai zasshi (May 1916), reprinted in Kushida Tamizō zenshū (Tokyo, 1935), I, 112.
30. "Kushida hōgakushi ni kotau," Kokka gakkai zasshi (June 1916), quoted by Furuta, p. 113. This article was included in Kawakami's Shakai mondai kanken, 1st ed. (Kyoto, 1918).
31. "Kawakami Hajime kyōju no Bimbō monogatari wo yomu," Kokka gakkai zasshi (April 1917), reprinted in Kushida Tamizō zenshū I, 289.

32. August 14, 1917, letter to Kushida, *Tegami*, p. 58.
33. February 4, 1918, letter to Kushida, *Tegami*, p. 62.
34. "Miketsukan," Osaka *Asahi shimbun*, January 1918, reprinted in *Chosakushū*, IX, 378.
35. Ibid., p. 381.
36. Shinobu Seisaburō, *Taishō seiji shi* (Tokyo, 1951), pp. 402–404. For a general discussion of the influence of the Russian Revolution on Japan, see Yamanabe Kentarō, "Jūgatsu kakumei ga Nihon ni ataeta eikyō," *Zen'ei* (December 1957), pp. 124–148.
37. Matsuo Takayoshi, *Taishō demokurashii no kenkyū* (Tokyo, 1966), p. 206.
38. "Miketsukan," *Chosakushū*, IX, 381.
39. "Seisan seisaku ka bumpai seisaku ka?" *Keizai ronsō* (May 1918), quoted by Furuta, p. 115.
40. Kawakami Hajime, "Beika mondai," Tokyo *Asahi shimbun*, August 26, 1918. Kawakami did not approve of the riots, which he felt were unmanly. He recommended that the rich be made, by legal means, to pay more taxes.
41. Yoshino Sakuzō who, like Kawakami, was a special contributor, also severed his connections with the newspaper over this incident. Soon afterward, Yoshino came to the defense of the *Asahi*, when it was attacked by an ultra-nationalist youth organization. See Takeda Kiyoko, "Yoshino Sakuzō," *Japan Quarterly* 12.4:520 (October–December 1965).
42. Watanabe Tōru, ed., *Kyōto chihō rōdō undō shi*, rev. ed. (Kyoto, 1968), p. 77.
43. Ibid., p. 84.
44. Sumiya Etsuji et al., *Nihon gakusei shakai undō shi* (Kyoto, 1953), p. 100.
45. Interview with Nishiyama Yūji, August 15, 1969. See also Kobayashi Terutsugi, "Omoitsuku mama," in Horie Muraichi, ed., *Kaisō no Kawakami Hajime*, p. 173.
46. Sumiya Etsuji, *Nihon gakusei shakai undō shi*, pp. 100–101.
47. "Zuihitsu (Dampen)," April 24, 1943. Reprinted in *Chosakushū*, IX, 330.
48. *Nihon gakusei shakai undō shi*, pp. 100–101.
49. "Aru isha no hitorigoto," *Shakai mondai kanken*, rev. ed., p. 400, originally published in Osaka *Asahi shimbun*, January 1, 1919.
50. "Omoide," quoted by Ōuchi Hyōe, "Kaidai," in Kawakami Hajime, *Bimbō monogatari* (Iwanami, 1961), p. 189.
51. *Tegami*, pp. 46–51.
52. Kawakami also wanted to continue his journalistic activities because he needed the additional income to pay for his son's medical bills and to support his parents in their old age. He almost agreed to write for *Nihon ichi*, an "extremely vulgar, popular magazine" sold at railroad newsstands, but Kushida and another friend, Kojima Yūma, talked him out of this idea. See Furuta Hikaru, *Kawakami Hajime*, p. 118.
53. Ibid. and *Jijoden*, pp. 134–135. See also Ōuchi Hyōe and Kojima Yūma, "Kawakami Hajime to Kushida Tamizō," in Kushida Tamizō, *Shakaishugi wa yami ni mensuru ka hikari ni mensuru ka* (Tokyo, 1951), supplement, pp. 299–300.

7. The Meaning of Marxism

1. *Jijoden*, p. 136.
2. "Dampen," April 24, 1943, *Chosakushū*, IX, 330.
3. December 30, 1924, letter to Kushida, *Tegami*, pp. 131–132.

4. Ibid., p. 133.
5. July 1, 1924, letter to Kushida, *Tegami,* p. 125.
6. Miyakawa Minoru, "Gakusha to shite no Kawakami sensei," in Horie Muraichi, ed., *Kaisō no Kawakami Hajime,* p. 237.
7. Ernst Fischer, ed., *The Essential Marx,* tr. Anna Bostock (New York, 1971), p. 133. "The working class," Marx wrote, in *The Poverty of Philosophy,* "in the course of its development, will substitute for the old civil society an association which will exclude classes and their antagonism."
8. Robert Tucker, *Philosophy and Myth in Karl Marx* (London, 1961), pp. 180–181.
9. "Socialism: Utopian and Scientific," in Lewis S. Feuer, ed., *Marx and Engels: Basic Writings on Politics and Philosophy* (New York, 1957), pp. 68–111.
10. *The Essential Marx,* p. 80.
11. Ibid., p. 126.
12. "Theses on Feuerbach," in Lewis S. Feuer, ed., *Marx and Engels: Basic Writings on Politics and Philosophy,* p. 245.
13. George Lichtheim, *Marxism: An Historical and Critical Study* (New York, 1961), p. 236.
14. The first ten issues of *Shakai mondai kenkyū,* hereafter cited as *Shaken,* were reprinted in *Kinsei keizai shisō shiron* (Kyoto, 1920). Excerpts from *Shaken* also appear in *Chosakushū,* vol. X.
15. *Kinsei keizai shisō shiron,* cited in Furuta Hikaru, *Kawakami Hajime,* p. 127.
16. *Yuibutsu shikan kenkyū* (Kyoto, 1921), p. 73. This is a collection of articles and translations on historical materialism appearing in *Keizai ronsō* and *Shaken* from 1919 to 1921.
17. "Shakaishugi no shinka," in *Chosakushū,* X, 156–157. This article was originally presented as a speech before the Osaka Industrial Trade Union Association on March 24, 1919 and afterward published in *Shaken* 5 (May 1919).
18. Ibid., pp. 158–163.
19. "Dampen," *Shaken* 2 (February 1919), reprinted in *Chosakushū,* VIII, 451.
20. "Shakaishugi shinka," *Chosakushū,* X, 165–166. Kawakami referred to Arthur Henderson's *The Aims of Labour* (London: Headley Brothers, 1918).
21. For Kawakami's lecture notes on the history of economic thought, see *Kinsei keizai shisō shiron* (1920) and *Shihonshugi keikaigaku no shiteki hatten* (1923). The second book was reprinted as vol. I of *Keizaigaku taikō* (1928), 2 vols. Both volumes appear in *Chosakushū,* vol. III.
22. Hiromi Ishigaki, "H. Kawakami's View on Bentham," *Hokudai Economic Papers* 1:74 (1968–69).
23. Kawakami Hajime, *Keizaigaku taikō,* vol. I, pt. 2, pp. 543 and 546 ff.
24. *Kinsei keizaigaku shiron,* cited in Frederick W. Poos, "Kawakami Hajime," Ph.D. diss. Stanford University, 1957, pp. 78–86.
25. *Keizaigaku taikō,* p. 629.
26. Hiromi Ishigaki, "H. Kawakami's View on Bentham," p. 63, from *Keizaigaku shiteki hatten, Chosakushū,* III, 387 and 551–553.
27. *Keizaigaku taikō,* pp. 859–869. Kawakami's treatment of Mill bears close resemblance to Robert L. Heilbroner's interpretation. See *The Wordly Philosophers* (New York, 1953), pp. 107–109.
28. See, for example, "Miru to Teirā fujin to no ren-ai," *Chosakushū,* IX, 403–407, originally published in *Warera* (August 1923); and "Shakaishugisha to shite no Zē Esu Miru," *Keizai ronsō* 8.4 (April 1919). Kawakami was inclined to treat utilitarian economics as part of the classical school of eco-

nomics; indeed, in his view, Bentham was the "writer who brought Classical Economics to great perfection." He thus neglected the reformist efforts that distinguished Bentham and his followers from the laissez-faire philosophy of Smith, Ricardo, and Malthus. See Hiromi Ishigaki, "H. Kawakami's View on Bentham," pp. 59–76.
29. Furuta Hikaru, *Kawakami Hajime*, p. 123.
30. Sakai Toshihiko, "Gendai shakaishugi no mottomo osorubeki kekkan," *Kaihō* 1 (June 1919), quoted by Furuta Hikaru, *Kawakami Hajime*, p. 121.
31. Werner Sombert, quoted by Robert Tucker, in *Philosophy and Myth in Karl Marx*, p. 12.
32. The word humanism, variously translated into Japanese as *jindōshugi*, *jimponshugi, jimbunshugi*, and *hyūmanizumu*, defies easy definition. In the sense of man-centered, as opposed to other-worldly, it may be used to describe societies as far apart as Confucian China and Renaissance Europe. In modern Japanese history, humanism tends to convey one of several possible meanings of individualism. When used by Japanese Christians in the Meiji period, for example, humanism connoted respect for the individual's inner, spiritual life and faith in man's potential for spiritual development. For the White Birch Society (Shirakaba-ha), humanism suggested something akin to artistic development—the encouragement of self-development and self-expression. In the Taishō period, under Kawakami's guidance, humanism took on the added meaning of social justice or humanitarianism. Japanese Marxists eventually gave the word a pejorative meaning—utopian, bourgeois, the ideology of the ruling class. Humanism thus became antithetical to Marxism, and in the postwar period, has become equated with socialist, but not Communist reform goals. Kawakami's thought contained aspects of all these definitions. It also included elements of traditional Confucian philanthropism—the responsibility of the elite to aid the poor—and Buddhist self-denial. For this reason, his brand of humanism was subject to the Marxist charge of being not only utopian and bourgeois but even feudal. As one author recently put it, Kawakami's humanism was "unconnected with modern consciousness." See Sumiya Etsuji et al., *Taishō demokurashii no shisō*, pp. 180–181. For a general discussion of Japanese humanism, see *Gendai Nihon Shisō Taikei*, vol. XXVII: *Hyūmanizumu*, ed. Odagiri Hideo (Tokyo, 1964).
33. Furuta Hikaru, *Kawakami Hajime*, pp. 121–122.
34. "Kahen no dōtoku to fukahen no dōtoku," *Shaken* 7 (July 1919), reprinted in *Chosakushū*, X, 180–183.
35. Kushida Tamizō, "Yuibutsu shikan to shakaishugi, Sakai-Kawakami nishi no roten," in *Kushida Tamizō zenshū*, ed. Ōuchi Hyōe (Tokyo, 1935), I, 7–17. Originally published in *Warera* (October 1919).
36. The book consisted of the first ten issues of *Shaken* as well as lectures on Adam Smith, Ricardo, and Malthus given in the spring of 1919.
37. "Marukusugaku ni okeru yuibutsu shikan no chii," in *Kushida Tamizō zenshū*, I, 33. This was Kushida's second review of Kawakami's book. The first, "Yuibutsu shikan to kaikyū tōsō-setsu oyobi seitō-ha keizaigaku to no kankei," was published in *Chosaku hyōron* (July 1920) and reprinted in *Kushida Tamizō zenshū*, vol. I. Kushida's review, according to Furuta Hikaru, represents the first time that Japanese Marxist research identified Marxist philosophy, i.e., dialectical materialism, as the basis of Marxist economics. See *Kawakami Hajime*, p. 129.
38. Richard Tucker, *Philosophy and Myth in Karl Marx*, pp. 197–198, from *The German Ideology*.
39. "Ningen no jiko manchakusei," *Shaken* 20 (November 1920), reprinted in *Chosakushū*, IX, 386.

40. June 8, 1920, letter to Kushida, *Tegami*, p. 72.

8. Historical Materialism and Revolutionary Will

1. Miyakawa Minoru, "Gakusha to shite no Kawakami Hajime," in Horie Muraichi, ed., *Kaisō no Kawakami Hajime* (Tokyo, 1948), p. 240. Toward the end of the nineteen twenties, Kawakami needed an entire week to prepare one issue.
2. Matsukata Saburō, "Taishō hachi-kyūnen goro no Kawakami Hajime," in Horie Muraichi, ed., *Kaisō no Kawakami Hajime*, pp. 208–209.
3. Ibid.
4. "Omoide," quoted by Ōuchi Hyōe, in Kawakami Hajime, *Bimbō monogatari* (Iwanami, 1946), p. 189.
5. "Shin-teki kaizō to butsu-teki kaizō," *Chosakushū*, X, 200–201, originally published in *Shaken* (March 1921). Kawakami's philosophical struggles with Marxism resemble the problems faced by the early Chinese Marxist, Li Ta-chao, who was influenced by Kawakami's pre-Marxist writings. Both men argued for spiritual as well as material reformation. Li's opposition to purely fatalistic interpretations of Marxism, however, was influenced by his desire for an immediate solution to China's political crisis and by his pre-Marxist commitment to political activism, whereas Kawakami's resistance to economic determinism was based on his pre-Marxist commitment to ethical activism. It took Kawakami many years before he translated his ethical consciousness into revolutionary practice. See Maurice Meisner, *Li Ta-chao and the Origins of Chinese Marxism* (Cambridge, 1967).
6. George Lichtheim, *Marxism: An Historical and Critical Study*, p. 240.
7. "Marukusu-setsu ni okeru shakai-teki kakumei to seiji-teki kakumei," *Chosakushū*, X, 254–258, originally published in *Shaken* (September 1922). Kawakami was referring to Tönnies' *Karl Marx* (1921) and Sombart's *Sozialismus und Soziale Bewegung* (1919).
8. "Marukusu ni okeru shakai-teki kakumei to seiji-teki kakumei," *Chosakushū* X, 266–268.
9. "Jiki shōso-naru shakai kakumei no kuwadate ni tsuite," *Keizai ronsō*, 15.4:513–514 (October 1922).
10. "Shakai kakumei to seiji kakumei," *Chosakushū*, X, 268–269.
11. "Marukusu no yuibutsu shikan ni kansuru ichi kōsatsu," *Keizai ronsō* 9.4 (July 1919). Reprinted in *Yuibutsu shikan kenkyū* (Kyoto, 1921). Italicized in the original.
12. "Shakai kakumei to seiji kakumei," pp. 268–275.
13. "Dampen," *Kaizō* (April 1921). This article, Kawakami's first censored work, touched off the "Toranamon Incident." See Chapter 9.
14. "Shakai kakumei to shakai seisaku," *Shaken* 38:1 (October 25, 1922). Reprinted in *Shakai soshiki to shakai kakumei* and in *Chosakushū*, vol. X.
15. Ernst Fischer, ed., *The Essential Marx* (New York, 1971), p. 87.
16. Isaiah Berlin, *Karl Marx* (New York, 1959), p. 146.
17. September 8, 1872, speech in Amsterdam, quoted by Bertram D. Wolfe, *Marxism* (New York, 1965), p. 214.
18. See, for example, Kawakami's "Marukusushugi no yuibutsu shikan ni kansuru ichi kōsatsu" and "Shakai kakumei no hitsuzensei to yuibutsu shikan," *Warera* 4 (May 1922).
19. "Dampen," *Chosakushū*, IX, 205, originally published in *Kaizō* (April 1921).
20. George Lichtheim, *Marxism: An Historical and Critical Study*, p. 240.

21. Inumaru Giichi, "Nihon marukusushugi no genryū," in Ikumi Takuichi, ed., *Nihon no marukusushugi,* Gendai no ideorogi (Tokyo, 1961), II, 8–12. Sakai's translation of Lenin's "The Revolution in Russia," in the October 1917 issue of *New Society,* was probably the first of Lenin's writings to be translated into Japanese. Takabatake Motoyuki, Sakai's associate, published the first article in Japan on Lenin's theories. Entitled "The Political Movement and the Economic Movement," it appeared in the February 1918 edition of *New Society.* In June 1919 Asō Hisashi, Kuroda Reiji, and Sano Manabu published *Kagekiha (The Bolsheviks),* the earliest complete introduction to the Russian Revolution appearing in Japan.
22. Kushida was in Europe between October 1920 and August 1922. Most of his time was spent in Berlin, studying and buying books on economics for the Ohara Social Problems Research Institute. On his return to Japan, he was placed in charge of editing the institute's magazine. See Ōuchi Hyōe, "Kushida Tamizō, Marukusugaku no kakuritsusha," *Asahi jyānaru* 5.1 (1963), reprinted in *Nihon no shisōka* (Tokyo, 1968), III, 150–167. Ōuchi considers Kushida the founder of academic Marxism in Japan.
23. "Omoide," p. 182.
24. The October 13, 1921 edition of *Shaken* contained that journal's first Russian work, Lenin's essay on the "Meaning of Agriculture Theory" ("Nōgyōsetsu no imi"). The essay is reprinted in *Shakai soshiki to shakai kakumei.*
25. "Yuibutsu shikan mondō—yuibutsu shikan to Roshiya kakumei," *Warera* 4 (January 1922); "Shakai kakumei no hitsuzensei to yuibutsu shikan," *Warera* 4 (May 1922); "Roshiya kakumei to shakaishugi kakumei," *Shaken* (January 1922), reprinted in *Chosakushū,* vol. X. This last article forms chap. 5 of *Shakai soshiki to shakai kakumei.*
26. "Marukusushugi ni iwayuru tokoro no kadoki ni tsuite," *Keizai ronsō* 13.6:939 (December 1921).
27. "Roshiya kakumei to shakaishugi kakumei," *Chosakushū,* X, 249, from Trotsky's *Our Revolution* (1918).
28. "Marukusushugi ni iwayuru tokoro no kadoki," pp. 932–933.
29. Ibid., pp. 932–934. "The words 'after the electrification of Russia,' " wrote Adam Ulam, "now assume in Lenin's mouth the same role that the words 'after the Revolution' played prior to October [1917]." See *The Bolsheviks* (New York, 1965), p. 481.
30. "Yuibutsu shikan mondō—yuibutsu shikan to Roshiya kakumei," *Warera* 4:46 (January 1922).
31. "Keizaigaku no kakumei," *Keizai ronsō* 15.2:305 (August 1922). "How the future will turn out is still a matter we cannot predict at the present time, but if labor-farmer Russia—and other countries—become socialist countries in the future, the natural result will be that the basic principle of economic policy will switch from laissez faire to management."
32. *Shihonshugi keizaigaku no shiteki hatten,* reprinted in *Keizaigaku taikō,* vol. I, pt. 2, pp. 874–876.
33. Kushida Tamizō, "Shakaishugi wa yami ni mensuru ka hikari ni mensuru ka?" *Kushida Tamizō zenshū,* I, 178–218, originally published in *Kaizō* (July 1924).
34. *Jijoden,* pp. 115–118.
35. Ibid. See also, Furuta Hikaru, *Kawakami Hajime,* pp. 136–140. Toward the end of his book, Kawakami quoted Ruskin's words, "Boldly raising the curtain, we face the light." These words reflected Ruskin's belief that it was possible to build a prosperous society on the basis of men's good side, not

their selfish, dark side. The title of Kushida's review thus asked the question, "Does socialism face toward darkness or light?" In the review, Kushida wrote, "Because socialism faces into the darkness, it gives birth to light," i.e., socialism confronted the dark side of society, instead of relying on naive optimism.

36. "Omoide," p. 182.
37. Jijoden, pp. 115–118, and June 16, 1924, letter to Kushida, Tegami, p. 114.
38. "Omoide," p. 181.
39. Ibid., translated in Ryūsaku Tsunoda, Theodore de Bary, and Donald Keene, Sources of the Japanese Tradition (New York, 1958), p. 821.

9. The Professor as Political Activist

1. Rodger Swearingen and Paul Langer, Red Flag in Japan (Cambridge, 1952), p. 49.
2. Miyakawa Minoru, "Gakusha to shite no Kawakami sensei," in Horie Muraichi, ed., Kaisō no Kawakami Hajime (Tokyo, 1948), pp. 240–241.
3. Fuji Tomio, Kyūdōsha, quoted in Nihon no hyakunen, VI, 183.
4. Matsuo Takayoshi, Taishō demokurashii no kenkyū (Tokyo, 1966), pp. 194 and 210.
5. Interview with Shiraishi Bon, August 20, 1969.
6. Matsukata Saburō, "Taishō hachi-kyūnen goro no Kawakami Hajime," in Horie Muraichi, ed., Kaisō no Kawakami Hajime, pp. 219–220.
7. Hasebe Fumio, "Deshi o megutte," in Horie Muraichi, ed., Kaisō no Kawakami Hajime, p. 253. A number of these students (Horie, Shiraishi, Miyakawa) came from Yamaguchi Prefecture, as did two brothers-in-law of Kawakami (Suekawa Hiroshi and Ōtsuka Yūshō) and several close friends (Sakuda Shōichi and Hososako Kanemitsu).
8. Horie Muraichi, "Kakan na jiko kakumei no rekishi," in Horie Muraichi, ed., Kaisō no Kawakami Hajime, p. 190.
9. Hasebe Fumio, "Deshi o megutte," p. 253.
10. Sumiya Etsuji, Nihon gakusei shakai undō shi (Kyoto, 1953), p. 102.
11. Tegami, pp. 79–80.
12. Hamura Shizuko, "Chichi no koto," Horie Muraichi, ed., Kaisō no Kawakami Hajime, p. 166.
13. Ōtsuka Yūshō, Mikan no tabiji, IV, 155.
14. The article, "Dampen," appeared in the April 1921 issue of the magazine Kaizō. Although this issue was immediately banned, the would-be assassin came across a copy of it two years later. For a brief description of the assassination attempt, see Nihon rekishi daijiten, XIV.
15. Sumiya Etsuji, Nihon gakusei shakai undō shi, pp. 181–183. For Kawakami's reaction to the incident, see his "Omoide," p. 332.
16. Miyakawa Minoru, "Gakusha to shite no Kawakami Hajime," p. 243.
17. Furuta Hikaru, Kawakami Hajime, p. 140.
18. Ibid., pp. 140–141.
19. Shakai kagaku kenkyū, p. 40, quoted by Fukumoto Kazuo, Yuibutsu shikan no tame ni (Tokyo, 1928), pp. 348–349.
20. Matsukata Saburō, "Taishō hachi-kyūnen goro no Kawakami Hajime," p. 238.
21. June 7, 1925, letter to Kushida Tamizō, Chosakushū, II, 384.
22. State and Revolution, postscript to 1st ed., quoted by Kawakami in Jijoden, p. 167.

23. Miyakawa Minoru, "Gakusha to shite no Kawakami Hajime," p. 233.
24. Isaiah Berlin, paraphrasing Marx's words, in *Karl Marx* (New York, 1959), p. 258.
25. "Is this not obscurantism, when pure theory is carefully partitioned off from practice," he asked, in *Materialism and Empirio-Criticism,* quoted by Nathan Leites, *A Study of Bolshevism* (Glencoe, 1953), p. 99.
26. Ibid., p. 218.
27. Gustav A. Wetter, *Dialectical Materialism* (New York, 1958), p. 125.
28. N. O. Lossky, *History of Russian Philosophy* (New York, 1951), p. 368.
29. "Dialectical Materialism," *Encyclopedia Brittanica* (1967 edition), VII, 357.
30. Wetter, *Dialectical Materialism,* p. 161.
31. Ibid., p. 125.
32. *Gokushū zeigo,* quoted by Sumiya Etsuji, "Kagaku-teki shinri to shūkyō-teki shinri no tōitsu," *Kiyō* 5:26 (March 1962).
33. Ibid., p. 14.
34. Sakuda Shōichi, *Jidai no hito: Kawakami Hajime,* p. 31.
35. Fukumoto Kazuo, *Yuibutsu shikan no tame ni,* p. 348.
36. July 22, 1925, letter to Kushida Tamizō, *Chosakushū,* IX, 387.
37. March 4, 1925, letter to Kushida Tamizō, *Tegami,* p. 135.
38. March 26, 1925, letter to Kushida Tamizō, *Tegami,* p. 139.
39. "Omoide," pp. 182–183.
40. Furuta Hikaru, *Kawakami Hajime,* pp. 144–145.
41. Miyakawa Minoru, "Gakusha to shite no Kawakami Hajime," pp. 243–244.
42. Furuta Hikaru, *Kawakami Hajime,* p. 146.
43. Ibid., p. 157. See also *Red Flag in Japan,* p. 21.
44. "Omoide," p. 185.
45. *Shakai kagaku kenkyū* 75 (November 1926), quoted by Furuta, p. 159.
46. Robert A. Scalapino, *The Japanese Communist Party* (Santa Monica, 1966), p. 46.
47. Ibid., p. 333.
48. Furuta Hikaru, *Kawakami Hajime,* p. 142.
49. *Shakai kagaku kenkyū* 77:25–26 (February 1927).
50. *Shakai kagaku kenkyū* 78:18 (March 1927).
51. *Jijoden,* pp. 152–153.
52. Letter to Kushida, *Jijoden,* p. 154.
53. *Red Flag in Japan,* pp. 29–30.
54. *Jijoden,* p. 154.
55. *Red Flag in Japan,* p. 30.
56. *Jijoden,* pp. 157–159.
57. Ibid., p. 243.
58. Ibid., p. 162.
59. Ibid., p. 143.
60. Personal interview with Kawakami Hide, autumn 1964.
61. Combined royalties from *An Outline of Economics (Keizaigaku taikō), Basic Theory of Marxist Economics (Marukusushugi keizaigaku no kiso riron),* and *Second Tale of Poverty (Daini bimbō monogatari)* amounted to over 39,000 yen, a very considerable amount of money in prewar Japan. See *Jijoden,* pp. 376–377.

10. Working for the Communist Party

1. *Jijoden,* pp. 168–169.
2. Ōuchi Hyōe and Kojima Yūma, "Kawakami Hajime to Kushida Tamizō," in Kushida Tamizō, *Shakaishugi wa yami ni mensuru ka hikari ni mensuru ka,* supplement, pp. 290–291.
3. Ibid., pp. 289–294.
4. *Jijoden,* pp. 166–167.
5. Ibid., p. 186.
6. Ibid., pp. 171–172.
7. Ibid., pp. 179–181; 186.
8. Ibid., pp. 182–186. The letter was written on December 24, 1928, in Moscow, but mailed from Frankfurt.
9. Ibid., pp. 187–188.
10. H. H. Gerth and C. Wright Mills, ed., *From Max Weber: Essays in Sociology* (New York, 1958), p. 121.
11. *Jijoden,* p. 64.
12. George O. Totten, *The Social Democratic Movement in Prewar Japan* (New Haven, 1966), p. 188.
13. Ōtsuka Yūshō, *Mikan no tabiji,* IV, 204.
14. Ibid., II, 40–41. For Kawakami's account of this incident, see *Jijoden,* p. 201.
15. Yamamoto was officially a representative of the Labor Farmer Federation for Securing Political Liberties, a group formed after the Labor Farmer party was banned in 1928. See Totten, *The Social Democratic Movement in Prewar Japan,* p. 201.
16. *Mikan no tabiji,* II, 56.
17. *Jijoden,* p. 202.
18. *Mikan no tabiji,* II, 57.
19. Tabei Kenji, *Ōyama Ikuo* (Tokyo, 1947), p. 77.
20. *Jijoden,* p. 305. When he moved to Tokyo to work for the New Labor Farmer party, Kawakami recalled, the Kyoto Seamen's Union assigned a seventeen year old youth to guard him. Otsuka's recollections are a little different. He remembered that two students took turns as bodyguards, living in the three-mat room on the second floor of the Kawakami house in Kyoto. They carried wooden swords as weapons and belonged to a student organization. One of them was Mizuta Mikio who, after the war, rose to a high position in the Liberal Democratic party. See *Mikan no tabiji,* II, 59.
21. Ibid., p. 334.
22. *Mikan no tabiji,* II, 45–46.
23. Ibid., p. 56.
24. Ibid., p. 46.
25. *Jijoden,* p. 339.
26. Ibid., pp. 221–224.
27. Ibid., p. 314–316.
28. *Jijoden,* cited by Furuta Hikaru, *Kawakami Hajime,* p. 182.
29. *Jijoden,* p. 245.
30. *Jijoden,* cited by Totten, *The Social Democratic Movement in Prewar Japan,* p. 166.
31. Ibid., p. 189.
32. Ibid., p. 164.

33. *Jijoden,* pp. 247–248.
34. Furuta Hikaru, *Kawakami Hajime,* pp. 182–183.
35. Letter to Hamura Nikio, May 11, 1930, in Furuta Hikaru, *Kawakami Hajime,* p. 183.
36. *Daini bimbō monogatari, Chosakushū,* II, 130–131.
37. *Marukusushugi keizaigaku no kiso riron* (Tokyo, 1929), I, 12.
38. *Daini bimbō monogatari, Chosakushū,* II, 240. "The liberation of the oppressed masses from exploitation—such a 'revolution' in social existence—can never be realized by the simple ethical theories of the book-worm."
39. Kawakami's major work on Marxist economics is *Marukusushugi keizaigaku no kiso riron,* vol. II.
40. *Jijoden,* p. 125.
41. *Keizaigaku taikō* (Tokyo, 1928), preface, p. 7.
42. *Jijoden,* pp. 298–299.
43. Ibid., pp. 333–337.
44. Ibid., pp. 310–313. He estimated that by September 1932 he had given over 20,000 yen to the Communist party.
45. *Red Flag in Japan,* pp. 37–40.
46. Furuta Hikaru, *Kawakami Hajime,* p. 187n.
47. *Jijoden,* pp. 332–333.
48. Furuta Hikaru, *Kawakami Hajime,* pp. 187–188.
49. Shiga Yoshio, "Kawakami hakase to kyōsantō," in Horie Muraichi, ed., *Kaisō no Kawakami Hajime,* pp. 79–82.
50. *Red Flag in Japan,* pp. 50–55.
51. *Jijoden,* pp. 344–345; 351.
52. Personal interview with Kawakami Hide, autumn 1964.
53. *Jijoden,* pp. 332, 341, 366.
54. Personal interview with Kawakami Hide, autumn 1964.
55. *Jijoden,* p. 371.
56. Ibid., p. 418. Iwata's death was reported in the *Yomiuri* newspaper on November 2, 1932. Kawakami Hide attended the funeral in her husband's place. Before his imprisonment, Iwata had tried to see Kawakami in order to convince him he was not a spy, but Kawakami had refused to see him, even though his fondness for Iwata, some seven years earlier, had been so great, he had given him his son's watch after the boy died.
57. Ibid., p. 373. Swearingen and Langer confirm the fact that Matsumura was working for the police. See, *Red Flag in Japan,* p. 55.
58. *Jijoden,* pp. 398–408.
59. Ibid., pp. 418 and 434. See also Ōtsuka Yūshō, *Mikan no tabiji,* III, 145–146.
60. Personal interview with Ōtsuka Yūshō, autumn 1964; *Jijoden,* p. 449; *Mikan no tabiji,* III, 155–172.
61. *Jijoden,* p. 453.
62. *Jijoden,* cited in Tsurumi Kazuko, *Social Change and the Individual* (Princeton, 1970), p. 74.
63. Quoted by Furuta, p. 189.
64. *Jijoden,* p. 351. Translation from Ryūsaku Tsunoda, William T. de Bary, and Donald Keene, *Sources of the Japanese Tradition,* p. 821.
65. Isaiah Berlin, *Karl Marx,* p. 272.
66. Ibid., pp. 154–155.

Epilogue

1. Robert A. Scalapino, *Democracy and the Party Movement in Prewar Japan* (Berkeley, 1962), p. 72.
2. Shihōshō, keiji kyoku, *Shisō jimu ni kansuru kunrei tsuchō shū, Shisō kenkyū shiryō tokushū* XXI, 178–179 (May 1935), quoted in Patricia G. Steinhoff, "Tenkō and Thought Control," paper presented at the Annual Meeting of the Association for Asian Studies, Boston, April 1–3, 1974, p. 10.
3. Sugiura Mimpei, "Kakumei jōnetsu o yusuburu," *Asahi jyānaru* 7.45:41 (October 1965).
4. Kawakami's younger daughter Yoshiko's two children lived with him and Hide. Yoshiko's husband, Suzuki Shigetoshi, had been employed by the Manchurian Railway in Dairen. When war broke out, he was drafted by the Japanese Army in Manchuria and shortly afterward, Yoshiko became seriously ill with a lung disease. Hide traveled to Dairen to help care for Yoshiko's children. About one and a half years later, she brought the children back with her to Kyoto. After Yoshiko's death, Hide continued to raise the grandchildren, until old age and rheumatism prevented her from caring for them any further. The younger child went to live in Tokyo with Shiraishi Bon, one of Kawakami's former students, and the older grandchild was taken by Kawakami's other daughter, Hamura Shizuko, who was married to a Kyoto Imperial University professor of engineering. Suzuki remained in China where he was a professor at Peking University. I am grateful to Shiraishi Bon for this information, communicated in a personal interview on August 20, 1969. Hamura Shizuko writes that Kawakami was far more affectionate with his grandchildren than with his children.
5. Frederick W. Poos, "Kawakami Hajime, 1879–1946: An Intellectual Biography," Ph.D. diss. Stanford University, 1957. Furuta Hikaru reports that, in the spring of 1964, on a trip to the Peoples' Republic of China, he saw Chinese translations of Kawakami's books in various libraries and bookstores. See his "Kawakami Hajime no shisō to sono mondaisei," *Kawakami Hajime chosakushū*, vol. X, supplement, p. 6.
6. Frederick Poos, "Kawakami Hajime," p. 29.
7. Five years was actually the minimum sentence for crimes in this category. Kawakami could have been sentenced to eight or nine years but, according to Ōya Sōichi, because the prosecuting attorney was also from Chōshū, Kawakami was treated more leniently. See "Tōsuiken no na no moto ni," *Asahi jyānaru*, 7.13:88 (March 28, 1965).
8. *Jijoden, Chosakushū*, VII, 81–82. Released from jail three years earlier, in 1934, Ōtsuka Yūshō went to work for the Manchurian Railway. At the end of the Pacific War, he surrendered to the Chinese Communist Eighth Route Army and worked for the Communists as a political adviser on Japanese politics. He also helped indoctrinate Japanese prisoners of war. He was eventually repatriated. Ōtsuka retained his membership in the Japanese Communist party, siding with the China faction when the party split. In the nineteen sixties, he was active in the Japan–China Friendship Society (Nichū-Yūkōkai), an organization seeking to establish greater cultural ties between China and Japan. Otsuka's contact in China was Suzuki Shigetoshi, a graduate of Kyoto Imperial University's economics department and the husband of Kawakami's younger daughter, Yoshiko.
9. Furuta Hikaru, *Kawakami Hajime*, p. 205, chronology; Sumiya Etsuji, *Kawakami Hajime*, pp. 334–335.
10. *Keizaigaku taikō*, introduction, p. 4. In his prison essays he wrote, "My faith in scholarship (my firm belief in the truth of Marxism) really did not move

one whit. However, this is perfectly natural, because a prison, unlike a university, does not have academic research facilities. Therefore, you cannot expect to be able to grasp scientific truth anew in such a place." See *Gokuchū zeigo* (Kyoto, 1947), pp. 20–21.
11. *Tōku de kasuka ni kane ga naru,* quoted by Fujita Shōzō, "Aru Marukusu shugisha," *Tenkō* (Tokyo, 1959), I, 242.
12. Persecution of Catholics in the sixteenth and early seventeenth centuries is an important exception.
13. Maruyama Masao, *Nihon no shisō* (Iwanami, 1961), pp. 55–59.
14. Ibid.
15. Robert A. Scalapino, "The Left Wing in Japan," *Survey* 43:107 (August 1962).
16. *Jijoden,* VI: 193.
17. Kawakami Tetsutarō, *Nihon no autosaidā,* p. 115.
18. Personal interview with Kawakami Hide, autumn 1964.
19. Kawakami Tetsutarō, *Nihon no autosaidā,* p. 116. In one of his poems, Kawakami wrote: "In autumn/the leaves will all have fallen/from the persimmon tree/near the gate of my home/where mother passes the day long, waiting."
20. Ōuchi Hyōe, *Takai yama: jimbutsu no arubamu* (Tokyo, 1963), p. 147.
21. *Jijoden,* quoted in Kazuko Tsurumi, *Social Change and the Individual,* p. 76.
22. Furuta Hikaru, *Kawakami Hajime,* chronology, pp. 253–255; Fujita Shōzō, "Aru Marukusu shugisha," *Tenkō,* 1, 244; *Jijoden, Chosakushū,* VII, 142–143, 178.
23. *Gokuchū zeigo, Chosakushū,* VIII, 377 ff. For an abridged translation, see Ryūsaku Tsunoda et al., *Sources of the Japanese Tradition* (New York, 1958), pp. 872–880.
24. Benjamin I. Schwartz, *Communist China and the Rise of Mao* (Cambridge, 1951), p. 23.
25. *Gokuchū zeigo,* quoted by Sumiya Etsuji, "Shakai kagaku-teki shinri to shūkyō-teki shinri no tōitsu," *Kiyō* 5.24 (1962).
26. This is the impression conveyed to me from talking with two of Kawakami's former students, Shiraishi Bon and Nishiyama Yūji. Both men were more eager to talk about the human appeal of Kawakami than about his Marxist thought and practice. Nishiyama keeps a portrait of Kawakami in his living room. Furuta Hikaru, Sumiya Etsuji, and Ōuchi Hyōe also stress the religious side of Kawakami's life, though Ōuchi especially thinks that Kawakami made important contributions to Japanese economic thought as well. I am deeply grateful to all five men for granting me personal interviews. See, too, Sugiura Mimpei's evaluation of the autobiography in "Kakumei jōnetsu o yusuburu," *Asahi jyānaru* 7.45:40–44 (October 1965), and Okuma Nobuyuki's treatment of Kawakami as a "Kyūdō no marukisuto," in *Nihon no shisōka* (Tokyo, 1968), III, 94–99.
27. At a meeting of the Kyoto Kawakami Commemorative Society in 1964 a portrait of Kawakami stood on the stage of the auditorium and *manjū,* Kawakami's favorite sweet, were sold to guests. Total membership in the Tokyo Commemorative Society in 1968 was 178. Among the members at that time were Shiraishi Bon, Nishiyama Yūji, Horie Muraichi, Suekawa Hiroshi, Amano Tetsujirō, Ōkuma Nobuyuki, Taniguchi Zentarō, and Miyagawa Minoru.
28. Ōkōchi Kazuo, "Kawakami Hajime to kyūdō no kagaku," *Chūō kōron* 3.439 (1965).
29. *Jijoden, Chosakushū,* VI, 227.

30. Personal interview with Ōtsuka Yūshō, autumn 1964.
31. Furuta Hikaru, *Kawakami Hajime*, p. 5, and Sugiura Mimpei, "Kakumei jōnetsu o yusuburu," *Asahi jyānaru* 7.45:43 (October 1965).
32. See, for example, *Jijoden, Chosakushū*, VI, 59–98. Elsewhere in his autobiography Kawakami gave as his reason for writing his remembrances the fact that "many great men—Rousseau and Kropotkin, Tolstoy and Gorky, Katayama Sen and Shimazaki Tōson—have all left remembrances of their early childhood. I should like to try to imitate them." (*Jijoden*, p. 5.)
33. Kazuko Tsurumi, *Social Change and the Individual*, p. 79.
34. Sumiya Etsuji, *Kawakami Hajime*, p. 307.

bibliography

A complete listing of everything published by Kawakami Hajime up to 1954 appears in Amano Keitarō, *Kawakami Hajime hakase bunkenshi*.

Kawakami's autobiographical writings constitute the single most important source of information on his life. These consist of several parts, including the early years, a self-portrait, and reminiscences. The autobiography was published in a five volume paperback edition by Iwanami *shoten* in 1952 and reissued by Chikuma shobō in 1964 as volumes VI and VII of the *Collected Works of Kawakami Hajime* (*Kawakami Hajime chosakushū*). Kawakami's prison diary (*Gokuchū nikki*), originally published in 1949, appears in volume XII of the *Collected Works*.

There are two full-length intellectual biographies, both entitled *Kawakami Hajime*. The one by Sumiya Etsuji is more detailed, while Furuta Hikaru's later study, which closely parallels Sumiya's work, is a more interpretive treatment of Kawakami's life and thought. Ōuchi Hyōe has written numerous lively accounts of Kawakami's life and times. His *Kawakami Hajime* (Tokyo: Chikuma shobō, 1966), is a series of essays on Kawakami's personality and thought.

A basic source for Kawakami's Marxist thought, in addition to the books and articles he published beginning in 1919, is his privately published journal, *Research in Social Problems* (*Shakai mondai kenkyū*). The first nine issues, all published in 1919, have been reprinted in Kawakami Hajime, *Shakai mondai kenkyū*, vol. I (Tokyo: Shakai shisōsha, 1975).

Most of Kawakami's major writings are contained in the twelve-volume *Collected Works*, though several important articles have been omitted and some of the larger works abridged. Unfortunately, there is no index to the twelve volumes, whose contents are only loosely arranged in chronological order. To remedy this deficiency, I have provided below a general guide to the *Collected Works*. In cases where the book or article I read in the original edition was reprinted in the *Collected Works*, I have tried to cite both references in the notes and bibliography.

Collected Works of Kawakami Hajime, by volume (Kawakami Hajime chosakushū)

I. Earliest economic writings, including *Shakaishugi hyōron* (A critique of socialism)

II. *Bimbō monogatari* (Tale of poverty); *Daini bimbō monogatari* (Second tale of poverty)
III. *Keizaigaku taikō* (An outline of economics)
IV. *Shihonron nyūmon* (A guide to Capital), vol. I
V. *Shihonron nyūmon*, vol. II
VI. *Jijoden* (Autobiography)
VII. *Jijoden* (Autobiography)
VIII. Economic writings at the end of the Meiji period, including *Jinsei no kisū* (Trends of life), *Keizai to jinsei* (Economics and human existence), *Jisei no hen* (Trends of the times)
"Gokuchū zeigo" (Leisurely meditation in prison)
"Aru oisha no hitorigoto" (Soliloquy of an unnamed doctor)
"Dampen" (Fragments)
IX. *Sokoku o kaerimite* (Reflections on our homeland)
"Omoide-Dampen" (Reminiscences-fragments)
"Miketsukan" (House of detention)
"Ningen no jiko manchakusei" (Man's self-deceiving nature)
"Kokindō zuihitsu" (Kokindō essays)
X. Kawakami's Writings in the Taishō period, including those immediately before and after his conversion to Marxism
Selections from *Shakai mondai kenkyū* (Research in social problems)
Marukusushugi no tetsugaku-teki kiso (The philosophical basis of Marxism)
XI. "Shiikashū" (Collection of Chinese and Japanese poetry)
"Shokanshū" (Collection of letters), including those previously unpublished
XII. *Gokuchū nikki* (Prison diary), excerpts
Bannen no seikatsu kiroku (Record of the later years), excerpts

Amano Keitarō 天野敬太郎. *Kawakami Hajime hakase bunkenshi* 河上肇博士文献志 (Published works of Dr. Kawakami Hajime). Tokyo: Nihon hyōronsha, 1956.
Analects of Confucius. Tr. Arthur Waley. New York: Vintage Press, 1938.
Balfour, F. H. "Court and Society in Tōkyō," *Transactions and Proceedings of The Japan Society, London* 3:54–74 (1893–94).
Brinkley, Frank, ed. *Japan Described and Illustrated by the Japanese*. 5 vols. Boston: J. B. Millet, 1897–1898.
Beasley, William. *The Meiji Restoration*. Stanford: Stanford University Press, 1972.
Bellah, Robert N. "Japan's Cultural Identity," *Journal of Asian Studies* 4:573–594 (August 1965).
―――― *Tokugawa Religion*. Glencoe: The Free Press, 1957.
Benedict, Ruth. *The Chrysanthemum and the Sword*. New York: World Publishing Company, 1972.

Bibliography

Berlin, Issiah. *Karl Marx: His Life and Environment.* New York: Oxford University Press, 1959.

Brown, Delmar. *Nationalism in Japan.* Berkeley: University of California Press, 1955.

Craig, Albert M. *Chōshū in the Meiji Restoration.* Cambridge: Harvard University Press, 1961.

DeBary, Wm. Theodore. *Self and Society in Ming Thought.* New York: Columbia University Press, 1970.

DeMartino, Richard. "The Human Situation and Zen Buddhism," in Suzuki, D. T., Erich Fromm, and Richard DeMartino, *Zen Buddhism and Psychoanalysis.* New York: Grove Press, 1960.

Dore, Ronald P. *Land Reform in Japan.* London: Oxford University Press, 1959.

Engels, Friederich. "Socialism: Utopian and Scientific," in Lewis S. Feuer, ed., *Marx and Engels: Basic Writings in Politics and Philosophy.* New York: Doubleday and Company, 1959.

Erikson, Erik. *Young Man Luther.* New York: W. W. Norton, 1962.

───── *Childhood and Society.* 2nd ed. New York: W. W. Norton, 1963.

Fischer, Ernst, ed. *The Essential Marx.* Tr. Anna Bostock. New York: Herder and Herder, 1971.

Fujita Shōzō 藤田省三. "Aru marukusu shugisha—Kawakami Hajime" 或るマルクス主義者,河上肇 (A Marxist—Kawakami Hajime), in Shisō no kagaku kenkyū kaiken, ed. *Tenkō* 転向 (Conversion). Vol. I. Tokyo: Heibonsha, 1959.

Fukui Kōji 福井孝治. "Ano koto, kono koto" あのことこのこと (This and that), in Horie Muraichi 堀江邑一, ed., *Kaisō no Kawakami Hajime* 回想の河上肇 (Kawakami Hajime remembered). Tokyo: Sekai hyōronsha, 1948.

Fukumoto Kazuo 福本和夫. *Yuibutsu shikan no tame ni* 唯物史観の為に (On behalf of historical materialism). Tokyo: Kaizōsha, 1928.

Furuta Hikaru 古田光. *Kawakami Hajime.* Tokyo: Tōkyō daigaku shuppankai, 1959.

───── "Kawakami Hajime no shisō to sono mondaisei" 河上肇の思想とその問題性 (Kawakami Hajime's thought and its problematical nature), *Kawakami Hajime chosakushū* 河上肇著作集 (Collected works of Kawakami Hajime). Vol. X, supplement. Tokyo: Chikuma shobō, 1964.

Gerth, H. H., and C. Wright Mills. *From Max Weber: Essays in Sociology.* New York: Oxford University Press, 1958.

Hall, Ivan P. *Mori Arinori.* Cambridge: Harvard University Press, 1973.

Hamura Shizuko 羽村しづ子. "Chichi no koto" 父のこと (About father). In Horie Muraichi, ed., *Kaisō no Kawakami Hajime* (Kawakami Hajime remembered). Tokyo: Sekai hyōronsha, 1948.

Hasebe Fumio 長谷部文雄. "Deshi o megutte" 弟子をめぐって (About his students). In Horie Muraichi, ed., *Kaisō no Kawakami Hajime* (Kawakami Hajime remembered). Tokyo: Sekai hyōronsha, 1948.

Hearn, Lafcadio. *Glimpses of Unfamiliar Japan*. Vol. I. Boston: Houghton Mifflin, 1894.

Heilbroner, Robert L. *The Worldly Philosophers*. Rev. ed. New York: Simon and Schuster, 1953.

Honjō Eijirō 本庄栄治郎. *Edo-Meiji jidai no keizai gakusha* 江戸明治時代の経済学者 (Economists of the Edo-Meiji period). Tokyo: Shibundō, 1962.

Horie Muraichi. "Kakan na jiko kakumei no rekishi" 果敢な自己革命の歴史 (The history of a bold personal revolution). In Horie Muraichi, ed., *Kaisō no Kawakami Hajime* (Kawakami Hajime remembered). Tokyo: Sekai hyōronsha, 1948.

Inumaru Giichi 犬丸義一. "Nihon marukusushugi no genryū" 日本マルクス主義の源流 (The origins of Japanese Marxism). In Ikumi Takuichi 井汲卓一, ed., *Nihon no marukusushugi* 日本のマルクス主義 (Japanese Marxism), *Gendai no ideorogī* 現代のイデオロギー (Modern ideology). Vol. II. Tokyo: San'ichi shobō, 1961.

Ishigaki Hiromi, "H. Kawakami's View on Bentham," *Hokudai Economic Papers* 1:59–76 (1968–1969).

Iwakuni shishi 岩国市史 (History of Iwakuni city). Iwakuni, 1957.

Jansen, Marius B., ed. *Changing Japanese Attitudes toward Modernization*. Princeton: Princeton University Press, 1965.

Japanese Society of London, Transactions and Proceedings, III (1893–94).

Kawakami Hajime. *Nihon sonnō ron* 日本尊農論 (On the reverence for agriculture in Japan). Tokyo: Yomiuri shimbun, 1905.

——— *Keizai gakujō no kompon kannen* 経済学上之根本観念 (Fundamental principles of political economics). Tokyo: Sendamoku sōsho, 1905. Reprinted in *Kawakami Hajime chosakushū* (Collected works of Kawakami Hajime). Vol. I. Tokyo: Chikuma shobō, 1964.

——— *Keizaigaku genron* 経済学原論 (Theories of political economics). Dōbunkan, 1905.

——— *Nihon nōseigaku* 日本農政学 (Agricultural management in Japan). Dōbunkan, 1906.

——— "Kakuhitsu no ji" 擱筆の辞 (A word on the end of these articles), *Shakaishugi hyōron* 社会主義評論 (Critique of socialism). 1st ed. Tokyo: Yomiuri shimbun, 1905. Reprinted in *Kawakami Hajime*, ed. Ōuchi Hyōe 大内兵衛. Gendai Nihon shisō taikei 現代日本思想大系 (Outline of modern Japanese thought). Vol. XIX. Tokyo: Chikuma shobō, 1964.

——— *Shakaishugi hyōron* (Critique of socialism). 1st ed. Tokyo: Yomiuri shimbun, 1906.

——— "Muga ai no shinri" 無我愛の真理 (The truth of selfless love). In *Kawakami Hajime*, ed. Ōuchi Hyōe. Gendai Nihon shisō taikei. Vol. XIX. Tokyo: Chikuma shobō, 1964.

——— *Jinsei no kisū* 人生の帰趣 (Trends in life). Tokyo: Yomiuri shimbun, 1906. Reprinted in *Kawakami Hajime chosakushū* (Collected works of

Kawakami Hajime). Vol. VIII. Tokyo: Chikuma Shobō, 1964.

——— "Keizai to dōtoku" 経済と道徳 (Economics and morality), *Nihon keizai shinshi* 日本経済新誌 (New magazine of Japanese economics) 1: 2–3 (April–May 1907). Reprinted in *Kawakami Hajime chosakushū* (Collected works of Kawakami Hajime). Vol. I. Tokyo: Chikuma shobō, 1964.

——— "Shūgi washo ni mirareta Kumazawa Banzan no keizai gakusetsu" 集義和書に見られたる熊沢蕃山の経済学説 (Economic theories of Kumazawa Banzan as seen in the *Shūgi washo*), *Kokka gakkai zasshi* 国家学会雑誌 (National Academic Society magazine) 21.10:1251–1263 (October 1907).

——— "Bakumatsu jidai no shakai shugisha Satō Nobuhiro" 幕末時代の社会主義者佐藤信淵 (Satō Nobuhiro, a socialist of the Bakumatsu period), *Kyoto hōgakkai zasshi* 京都法学会雑誌 (Kyoto law society magazine), 4:10 (October 1909).

——— *Keizai to jinsei* 経済と人生 (Economics and human existence). Tokyo: Jitsugyō no Nihonsha, 1911. Reprinted in *Kawakami Hajime chosakushū* (Collected works of Kawakami Hajime). Vol. VIII. Tokyo: Chikuma shobō, 1964.

——— *Jisei no hen* 時勢之変 (Trends of the times). Tokyo: Yomiuri shimbun, 1911. Reprinted in *Kawakami Hajime chosakushū* (Collected works of Kawakami Hajime). Vol. VIII. Tokyo: Chikuma shobō, 1964.

——— "Dāuinizumu to Marukishizumu" ダーウィニズムとマルキシズム (Darwinism and Marxism), *Chūō kōron* 中央公論 (Central forum), 27.4:16–37 (April 1, 1912).

——— *Keizaigaku kenkyū* 経済学研究 (Studies in economics). Kyōritsusha, 1912. Reprinted in *Kawakami Hajime chosakushū* (Collected works of Kawakami Hajime). Vol. I. Tokyo: Chikuma shobō, 1964.

——— *Sokoku o kaerimite* 祖国を顧みて (Reflections on our homeland). Tokyo: Jitsugyō no Nihonsha, 1915.

——— "Sumāto no 'Ichi keizai gakusha no daini shisō'" スマアトの「一経済学者の第二思想」 (Smart's "Second thoughts of an economist"), *Keizai ronsō* 経済論叢 (Collection of economic treatises) 4:2 (February 1917). Reprinted in *Kawakami Hajime chosakushū* (Collected works of Kawakami Hajime). Vol. X. Tokyo: Chikuma shobō, 1964.

——— *Bimbō monogatari* 貧乏物語 (Tale of poverty). Kyoto: Kōbundō, 1917. Reprinted in *Kawakami Hajime chosakushū* (Collected works of Kawakami Hajime). Vol. II. Tokyo: Chikuma shobō, 1964.

——— "Miketsukan" 未決監 (House of detention). *Osaka Asahi shimbun* 大阪朝日新聞 (Osaka Asahi newspaper), January 1918. Reprinted in *Kawakami Hajime chosakushū* (Collected works of Kawakami Hajime). Vol. IX. Tokyo: Chikuma shobō, 1964.

——— "Beika mondai" 米価問題 (The problem of rice prices), *Tokyo Asahi shimbun* 東京朝日新聞 (Tokyo Asahi newspaper), August 19–26, 1918.

―――― *Shakai mondai kanken* 社会問題管見 (Current views on social problems). Kyoto: Kōbundō, 1918.

―――― "Aru isha no hitorigoto" 或医者の独語 (Soliloquy of an unnamed doctor), *Osaka Asahi shimbun* (Osaka Asahi newspaper), January 1, 1919. Reprinted in Kawakami Hajime, *Shakai mondai kanken* (Current views on social problems). Rev. ed. Kyoto: Kōbundō, 1920.

―――― "Dampen" 断片 (Fragments), *Shakai mondai kenkyū* 社会問題研究 (Research in social problems) 2 (February 1919). Reprinted in *Kawakami Hajime chosakushū* (Collected works of Kawakami Hajime). Vol. VIII. Tokyo: Chikuma shobō, 1964.

―――― "Shakaishugisha to shite no Zē Esu Miru" 社会主義者としてのゼー・エス・ミル (J. S. Mill as a socialist), *Keizai ronsō* (Collection of economic treatises) 8.4:30–37 (April 1919).

―――― "Shakaishugi no shinka" 社会主義の進化 (The evolution of socialism), *Shakai mondai kenkyū* (Research in social problems) 5 (May 1919). Reprinted in *Kawakami Hajime chosakushū* (Collected works of Kawakami Hajime). Vol. X. Tokyo: Chikuma shobō, 1964.

―――― "Marukusu no yuibutsu shikan ni kansuru ichi kōsatsu" マルクスの唯物史観に関する一考察 (An inquiry into Marx's historical materialism), *Keizai ronsō* (A collection of economic treatises) 9.4:1–17 (July 1919). Reprinted in Kawakami Hajime, *Yuibutsu shikan kenkyū* 唯物史観研究 (Studies in historical materialism). Kyoto: Kōbundō, 1921.

―――― "Kahen no dōtoku to fukahen no dōtoku" 可変の道徳と不可変の道徳 (Changing morality and unchanging morality), *Shakai mondai kenkyū* (Research in social problems) 7 (July 1919). Reprinted in *Kawakami Hajime chosakushū* (Collected works of Kawakami Hajime). Vol. X. Tokyo: Chikuma shobō, 1964.

―――― *Kinsei keizai shisō shiron* 近世経済思想史論 (History of modern economic thought). Kyoto: Kōbundō, 1920.

―――― "Ningen no jiko manchakusei" 人間の自己瞞着性 (Man's self-deceiving nature), *Shakai mondai kenkyū* 20 (November 1920). Reprinted in *Kawakami Hajime chosakushū* (Collected works of Kawakami Hajime). Vol. IX. Tokyo: Chikuma shobō, 1964.

―――― "Shin-teki kaizō to butsu-teki kaizō" 心的改造と物的改造 (Spiritual reconstruction and material reconstruction), *Shakai mondai kenkyū* (Research in social problems) 21 (March 1921). Reprinted in *Kawakami Hajime chosakushū* (Collected works of Kawakami Hajime). Vol. X. Tokyo: Chikuma shobō, 1964.

―――― "Dampen" (Fragments), *Kaizō* 改造 (Liberation) 3: 4 (April 1, 1921). Reprinted in *Kawakami Hajime chosakushū* (Collected works of Kawakami Hajime). Vol. IX. Tokyo: Chikuma shobō, 1964.

―――― "Marukusushugi ni yū tokoro no katoki ni tsuite" マルクス主義に言う所の過渡期に就いて (On the meaning of the "transition period" in

Marx), *Keizai ronsō* (Collection of economic treatises) 13.6: 150–157 (December 1921).

——— *Yuibutsu shikan kenkyū* (Studies in historical materialism). Kyoto: Kōbundō, 1921.

——— "Yuibutsu shikan mondō—yuibutsu shikan to Roshiya kakumei" 唯物史観問答—唯物史観と露西亜革命 (A dialogue on historical materialism—historical materialism and the Russian Revolution), *Warera* 我等 (We) 4.1: 41–47 (January 1, 1922).

——— "Roshiya kakumei to shakaishugi kakumei" 露西亜革命と社会主義革命 (The Russian revolution and socialist revolution), *Shakai mondai kenkyū* 29 (January 18, 1922). Reprinted in *Kawakami Hajime chosakushū* (Collected works of Kawakami Hajime). Vol. X. Tokyo: Chikuma shobō, 1964.

——— "Shakaishugi kakumei no hitsuzensei to yuibutsu shikan" 社会主義革命の必然性と唯物史観 (The necessity of socialist revolution and historical materialism), *Warera* (We) 4.5: 21–27 (May 1, 1922).

——— "Keizaigaku no kakumei" 経済学の革命 (The revolution in economics), *Keizai ronsō* (Collection of economic treatises) 15.2: 155–160 (August 1922).

——— "Marukusu-setsu ni okeru shakai-teki kakumei to seiji-teki kakumei" マルクス説における社会的革命と政治的革命 (Social and political revolution in Marx's theories), *Shakai mondai kenkyū* (Research in social problems) 37 (September 13, 1922). Reprinted in *Kawakami Hajime chosakushū* (Collected works of Kawakami Hajime). Vol. X. Tokyo: Chikuma shobō, 1964.

——— "Shakai kakumei to shakai seisaku" 社会革命と社会政策 (Social revolution and social policy), *Shakai mondai kenkyū* (Research in social problems) 38 (October 25, 1922). Reprinted in *Kawakami Hajime chosakushū* (Collected works of Kawakami Hajime). Vol. X. Tokyo: Chikuma shobō, 1964.

——— "Jiki shōsō-naru shakai kakumei no kuwadate ni tsuite" 時機尚早なる社会革命の企について (Plans for a premature social revolution), *Keizai ronsō* (Collection of economic treatises) 15: 4 (October 1922). Reprinted in Kawakami Hajime, *Shakai soshiki to shakai kakumei* 社会組織と社会革命 (Social organization and social revolution). Kyoto: Kōbundō, 1922.

——— *Shakai soshiki to shakai kakumei* (Social organization and social revolution). Kyoto: Kōbundō, 1922.

——— "Miru to Teirā fujin to no ren-ai" ミルとテイラア夫人との恋愛 (The love affair between Mill and Mrs. Taylor), *Warera* (We) 5: 8 (August 1923). Reprinted in *Kawakami Hajime chosakushū* (Collected works of Kawakami Hajime). Vol. X. Tokyo: Chikuma shobō, 1964.

——— *Shihonshugi keizaigaku no shiteki hatten* 資本主義経済学の史的発展 (Historical development of capitalist economics). Kyoto: Kōbundō, 1923.

——— "Sobo no tsuioku" 祖母の追憶 (Remembrances of grandmother), *Warera* (We) 6: 1 (August 1923). Reprinted in *Kawakami Hajime chosakushū*

(Collected works of Kawakami Hajime). Vol. VIII. Tokyo: Chikuma shobō, 1964.

——— "Yuibutsu shikan ni kansuru jiko seisan" 唯物史観に関する自己清算 (A personal settlement of accounts with historical materialism), *Shakai mondai kenkyū* (Research in social problems), pts. 1–3, vols. 77–79 (February–April 1927); pts. 4–7, vols. 80–82 (June–August 1927).

——— *Keizaigaku taikō* 経済学大綱 (Fundamental principles of economics). 2 vols. Tokyo: Kaizōsha, 1928.

——— *Marukusushugi keizaigaku no kiso riron* マルクス主義経済学の基礎理論 (Basic theory of Marxist economics). Tokyo: Kaizōsha, 1929.

——— *Daini bimbō monogatari* 第二貧乏物語 (The second tale of poverty). Tokyo: Kaizōsha, 1930. Reprinted in *Kawakami Hajime chosakushū* (Collected works of Kawakami Hajime). Vol. II. Tokyo: Chikuma shobō, 1964.

——— "Zuihitsu (Dampen)" 随筆「断片」(Miscellany [Fragments]), April 24, 1943. Reprinted in *Kawakami Hajime chosakushū* (Collected works of Kawakami Hajime). Vol. IX. Tokyo: Chikuma shobō, 1964.

——— *Gokuchū zeigo* 獄中贅語 (Leisurely meditation in prison). Kyoto: Kawara shoten, 1947.

——— *Bimbō monogatari* (Tale of poverty). Tokyo: Iwanami shoten, 1961.

——— "Omoide" 思い出 (Reminiscences). In *Kawakami Hajime chosakushū* (Collected works of Kawakami Hajime). Vol. IX. Tokyo: Chikuma shobō, 1964.

Kawakami Hide 河上秀. *Rusu nikki* 留守日記 (A diary in his absence). Tokyo: Chikuma shobō, 1967.

Kawakami Tetsutarō 河上徹太郎. *Nihon no autosaidā* 日本のアウトサイダー (Japanese outsider). Tokyo: Chūō kōron, 1961.

Kishimoto Hideo. *Japanese Religion in the Meiji Era*. Tr. John F. Howes. Tokyo: Ōbunsha, 1956.

Kobayashi Terutsugu 小林輝次. "Omoitsuku mama" 思いつくまま (Miscellaneous thoughts), in Horie Muraichi, ed., *Kaisō no Kawakami Hajime* (Kawakami Hajime remembered). Tokyo: Sekai hyōronsha, 1948.

Kosaka Masaaki. *Japanese Culture in the Meiji Era*. Tr. David Abosch. Tokyo: Pan-Pacific Press, 1958.

Kublin, Hyman. *Asian Revolutionary: The Life of Katayama Sen*. Princeton: Princeton University Press, 1964.

Kushida Tamizō 櫛田民蔵. "Kawakami kyōju no 'Shashi to hinkon' o yomite" 河上教授の「奢侈と貧困」を読みて (Reading Professor Kawakami's "Luxury and Poverty"). In *Yuibutsu shikan* 唯物史観 (Historical materialism), ed. Ōuchi Hyōe. *Kushida Tamizō zenshū* 櫛田民蔵全集 (Complete works of Kushida Tamizō). Vol. I. Tokyo: Kaizōsha, 1935.

——— "Kawakami Hajime kyōju no *Bimbō monogatari* o yomu" 河上肇教授の「貧乏物語」を読む (Reading Professor Kawakami Hajime's *Tale of Poverty*). In *Yuibutsu shikan* (Historical materialism), ed. Ōuchi Hyōe.

Kushida Tamizō zenshū (Complete works of Kushida Tamizō). Vol. I. Tokyo: Kaizōsha, 1935.

———. "Yuibutsu shikan to shakaishugi, Sakai-Kawakami nishi no ronten" 唯物史観と社会主義, 堺, 河上二氏の論点 (Historical materialism and socialism, the points at issue between Sakai and Kawakami), *Warera* (We), October 1919. Reprinted in *Yuibutsu shikan* (Historical materialism), ed. Ōuchi Hyōe. *Kushida Tamizō zenshū*. Vol. I. Tokyo: Kaizōsha, 1935.

———. "Marukusugaku ni okeru yuibutsu shikan no chii" マルクス学における唯物史観の地位 (The position of historical materialism within Marxism). In *Yuibutsu shikan* (Historical materialism), ed. Ōuchi Hyōe. *Kushida Tamizō zenshū* (Complete works of Kushida Tamizō). Vol. I. Tokyo: Kaizōsha, 1935.

———. "Shakaishugi wa yami ni mensuru ka hikari ni mensuru ka?" 社会主義は闇に面するか光に面するか (Does socialism face toward darkness or toward light?). In *Yuibutsu shikan* (Historical materialism), ed. Ōuchi Hyōe. *Kushida Tamizō zenshū* (Complete works of Kushida Tamizō). Vol. I. Tokyo: Kaizōsha, 1935.

Leites, Nathan. *A Study of Bolshevism*. Glencoe: The Free Press, 1953.

Lichtheim, George. *Marxism: An Historical and Critical Study*. New York: Praeger, 1961.

Lifton, Robert Jay. "The Struggle for Cultural Rebirth," *Harpers Magazine*, pp. 84-90 (April 1973).

Lossky, N. O. *History of Russian Philosophy*. New York: International Universities Press, 1951.

Maruyama Masao 丸山真男. *Nihon no shisō* 日本の思想 (Japanese thought). Tokyo: Iwanami, 1961.

———. "Patterns of Individuation and the Case of Japan: A Conceptual Scheme," in Marius Jansen, ed., *Changing Japanese Attitudes toward Modernization*. Princeton: Princeton University Press, 1965.

Marx, Karl. "Theses on Feuerbach," in Lewis S. Feuer, ed., *Marx and Engels: Basic Writings on Politics and Philosophy*. New York: Doubleday and Company, 1959.

Matsukata Saburō 松方三郎. Taishō hakkunen goro no Kawakami Hajime" 大正八・九年頃の河上肇 (Kawakami Hajime around 1919-1920). In Horie Muraichi, ed., *Kaisō no Kawakami Hajime* (Kawakami Hajime remembered) Tokyo: Sekai hyōronsha, 1948.

Matsuo Takayoshi 松尾尊允. *Taishō demokurashii no kenkyū*. 大正デモクラシーの研究 (Studies on Taishō democracy). Tokyo: Aoki shoten, 1966.

Meisner, Maurice. *Li Ta-chao and the Origins of Chinese Marxism*. Cambridge: Harvard University Press, 1967.

Mitchell, Wesley C. *Lecture Notes on Types of Economic Theory*. New York: Augustus M. Kelley, 1949.

Miyakawa Minoru 宮川實. "Gakusha to shite no Kawakami Hajime" 学者

としての河上肇 (Kawakami Hajime as a scholar). In Horie Muraichi, ed., *Kaisō no Kawakami Hajime* (Kawakami Hajime remembered). Tokyo: Sekai hyōronsha, 1948.

Nihon no hyakunen 日本の百年 (A century of Japan). *Narikin tenka* 成金天下 (Nouveaux riches society). Vol. VI. Tokyo: Chikuma shobō, 1962.

Odagiri Hideo 小田切秀雄, ed, *Hyūmanizumu* ヒューマニズム (Humanism). Gendai Nihon shisō taikei (Outline of modern Japanese thought). Vol. XVII. Tokyo: Chikuma shobō, 1964.

Ohkawa, Kazushi, and Henry Rosovsky. "A Century of Japanese Growth." In William W. Lockwood, ed., *The State and Japanese Enterprise in Japan*. Princeton: Princeton University Press, 1965.

Ōkochi Kazuo 大河内一男, ed. *Shakaishugi* 社会主義 (Socialism). Gendai Nihon shisō taikei (Outline of modern Japanese thought). Vol. XVI. Tokyo: Chikuma shobō, 1963.

────── "Kawakami Hajime to kyūdō no kagaku" 河上肇と求道の科学 (Kawakami Hajime and the science of the search for the Way), *Chūō kōron* (Central forum) 3: 433–439 (March 1965).

Ōkuma Nobuyuki 大熊信行. "Kawakami Hajime, kyūdō no marukisuto" 河上肇求道のマルキスト (Kawakami Hajime, a Marxist in search of the Way). In *Nihon no shisōka* 日本の思想家 (Japanese thinkers). Vol. III. Tokyo: Asahi shimbun, 1968.

Olson, Lawrence. *Dimensions of Japan*. New York: American Universities Field Staff, 1963.

Ōya Sōichi 大宅壮一. "Tōsuiken no na no moto ni" 総帥権の名のもとに (Under the name of the authority of the supreme command), *Asahi jyānaru* 朝日ジャーナル (Asahi journal) 7.13: 84–90 (March 28, 1965).

Ōtsuka Yūshō 大塚有章. *Mikan no tabiji* 未完の旅路 (Unfinished journey). 4 vols. Tokyo: San'ichi shobō, 1961.

Ōuchi Hyōe, ed. *Kawakami Hajime yori Kushida Tamizō e no tegami* 河上肇より櫛田民蔵への手紙 (Letters from Kawakami Hajime to Kushida Tamizō). Tokyo: Kamakura Bunko, 1947.

────── *Keizaigaku gojūnen* 経済学五十年 (Fifty years of economics). Tokyo: Tokyo daigaku shuppankai, 1959.

────── *Takai yama: jimbutsu arubamu* 高い山: 人物アルバム (High mountains: an album of personalities). Tokyo: Iwanami, 1963.

──────, ed. *Kawakami Hajime*. Gendai Nihon shisō taikei (An outline of modern Japanese thought). Vol. XIX. Tokyo: Chikuma shobō, 1964.

────── "Keizai gakusha to shite no Kawakami Hajime" 経済学者としての河上肇 (Kawakami Hajime as an economist). In Suekawa Hiroshi 末川博, ed., *Kawakami Hajime no kenkyū* 河上肇の研究. (Studies on Kawakami Hajime). Tokyo: Chikuma shobō, 1965.

────── *Kawakami Hajime*. Tokyo: Chikuma shobō, 1966.

────── "Kushida Tamizō, Marukusugaku no kakuritsusha" 櫛田民蔵: マル

クス学の確立者 (Kushida Tamizō, the founder of Marxist studies), *Nihon no shisōka* (Japanese thinkers). Vol. III. Tokyo: Asahi shimbun, 1968.

——— and Kojima Yūma 小島祐馬. "Kawakami Hajime to Kushida Tamizō" 河上肇と櫛田民蔵 (Kawakami Hajime and Kushida Tamizō). In Kushida Tamizō, *Shakaishugi wa yami ni mensuru ka hikari ni mensuru ka* (Does socialism face toward darkness or toward light?). Supplement. Tokyo: Asahi bunko, 1951.

Poos, Frederich W. "Kawakami Hajime (1879–1946): An Intellectual Biography." Ph. D. diss. Stanford University, 1957.

Pritchett, V. S. "The Strength of an Injured Spirit," *New York Review of Books* (November 2, 1972).

Rōdō sekai 労働世界 (Labor World) 6: 15 (September 13, 1902).

Sakuda Shōichi 作田荘一. *Jidai no hito: Kawakami Hajime* 時代の人: 河上肇 (A man of the times: Kawakami Hajime). Osaka: Kaikensha, 1949.

Scalapino, Robert A., "The Left Wing in Japan," *Survey* 43: 102–111 (August 1962).

——— *Democracy and the Party Movement in Prewar Japan*. Berkeley: University of California Press, 1962.

——— *The Japanese Communist Movement, 1920–1965*. Santa Monica: Rand Corporation, 1966.

Schumpeter, Joseph. *History of Economic Analysis*. New York: Oxford University Press, 1954.

Schwartz, Benjamin I. *Communist China and the Rise of Mao*. Cambridge: Harvard University Press, 1951.

Seligman, Edwin R. A. *The Economic Interpretation of History*. 2nd. ed. New York: Columbia University Press, 1907.

Shibusawa Keizō. *Japanese Life and Culture in the Meiji Era*. Tr. Charles S. Terry. Tokyo: Ōbunsha, 1958.

——— *Japanese Society in the Meiji Era*. Tr. Aora H. Culbertson and Michiko Kimura. Tokyo: Ōbunsha, 1958.

Shiga Yoshio 志賀義雄. "Kawakami hakase to kyōsantō" 河上博士と共産党 (Dr. Kawakami and the Communist party). In Horie Muraichi, ed., *Kaisō no Kawakami Hajime* (Kawakami Hajime remembered). Tokyo: Sekai hyōronsha, 1948.

Shimazaki Tōson 島崎藤村. "Etoranzee shō" エトランゼエ抄 (A stranger's essay). In Suekawa Hiroshi, ed., *Kawakami Hajime kenkyū* (Studies on Kawakami Hajime). Tokyo: Chikuma shobō, 1965.

Shinobu Seisaburō 信夫清三郎. *Taishō seiji shi* 大正政治史 (A history of Taishō politics). Tokyo: Kawade shobō, 1951.

Smith, Henry D., II. *Japan's First Student Radicals*. Cambridge: Harvard University Press, 1972.

Steinhoff, Patricia G. "Tenkō: Ideology and Social Integration in Prewar Japan." Ph. D. diss. Harvard University, 1969.

——— "Tenkō and Thought Control." Paper presented at the Annual Meeting of the Association for Asian Studies, Boston, April 1–3, 1974.

Sugiura Mimpei 杉浦明平. "Kakumei jōnetsu o yusuburu" 革命情熱を揺ぶる (Arousing revolutionary ardor), *Asahi jyānaru* 7.45: 40–44 (October 1965).

Sumiya Etsuji 住谷悦治. *Nihon keizaigaku shi no hitokoma* 日本経済学史の一齣 (A phase of the history of Japanese economics). Tokyo: Nihon hyōronsha, 1948.

——— *Nihon keizaigaku shi* 日本経済学史 (History of Japanese economics). Kyoto: Mineruba shobō, 1958.

——— "Kagaku-teki shinri to shūkyō-teki shinri no tōitsu" 科学的真理と宗教的真理の統一 (The unity of scientific truth and religious truth), *Kiyō* 紀要 (Bulletin). Vol. V. Kyoto: Dōshisha daigaku, 1962.

——— *Kawakami Hajime*. Tokyo: Yoshikawa kōbunkan, 1962.

———, Takakuwa Suehide 高桑末秀, and Ogura Jōji 小倉襄二. *Nihon gakusei shakai undō shi* 日本学生社会運動史 (History of the Japanese left-wing student movement). Kyoto: Dōshisha daigaku, 1953.

———, Yamaguchi Kōsaku 山口光朔, Koyama Hitoshi 小山仁示, Asada Mitsuteru 浅田光輝, and Koyama Hirotake 小山弘健. *Taishō demokurashii no shisō* 大正デモクラシーの思想 (Taishō democratic thought). Tokyo: Chikuma shobō, 1967.

Suzuki, D. T., Erich Fromm, and Richard DeMartino. *Zen Buddhism and Psychoanalysis*. New York: Grove Press, 1960.

Swearingen, Rodger, and Paul Langer. *Red Flag in Japan*. Cambridge: Harvard University Press, 1952.

Tabei Kenji 田部井健次. *Ōyama Ikuo* 大山郁夫. Tokyo: Shinrosha, 1947.

Takeda Akira 竹田省. "Ōshū ryūgaku jidai no Kawakami-san" 欧州留学時代の河上さん (Mr. Kawakami in his European overseas student days). In Horie Muraichi, ed., *Kaisō no Kawakami Hajime* (Kawakami Hajime remembered). Tokyo: Sekai hyōronsha, 1948.

Takeda Kiyoko. "Yoshino Sakuzō," *Japan Quarterly* 12.4: 515–524 (October–December 1965).

Tokutomi Iichirō 徳富猪一郎. *Kōshaku Yamagata Aritomo den* 公爵山県有朋伝 (Biography of Prince Yamagata Aritomo). 3 vols. Tokyo: Hara shobō, 1969. Reprint of 1933 ed.

Tolstoy, Leo. *My Religion*. Tr. Leo Wiener. *The Complete Works of Count Tolstoy*. Vol. XVI. London: J. M. Dent, 1904.

Totten, George O., III. *The Social Democratic Movement in Prewar Japan*. New Haven: Yale University Press, 1966.

Trotsky, Leon. *Our Revolution*. Tr. Moissaye J. Olgin. New York: Henry Holt and Company, 1918.

Tsuda Seifū 津田青楓. "Kawakami hakase no ningensei" 河上博士の人間性 (Dr. Kawakami's humanity). In Suekawa Hiroshi, ed., *Kawakami Hajime kenkyū* (Studies on Kawakami Hajime). Tokyo: Chikuma shobō, 1965.

Tsunoda, Ryusaku, William DeBary and Donald Keene. *Sources of the Japanese Tradition*. New York: Columbia University Press, 1958.

Tsurumi, Kazuko. *Social Change and the Individual: Japan before and after Defeat in World War II*. Princeton: Princeton University Press, 1970.

Tucker, Robert. *Philosophy and Myth in Karl Marx*. London: Cambridge University Press, 1961.

Uchida Yoshihiko 内田義彦. "Meiji makki no Kawakami Hajime" 明治末期の河上肇 (Kawakami Hajime at the end of the Meiji period). In *Nihon shihonshugi no sho mondai* 日本資本主義の諸問題 (Some problems in Japanese capitalism), ed. Yamada Moritarō 山田盛太郎. Tokyo: Miraisha, 1960.

Ulam, Adam B. *The Bolsheviks*. New York: MacMillan Company, 1965.

Vogel, Ezra. *Japan's New Middle Class*. Berkeley: University of California Press, 1963.

Watanabe Tōru 渡部徹, ed. *Kyōto chihō rōdō undō shi* 京都地方労働運動史 (A history of the labor movement in the Kyoto region). Rev. ed. Kyoto: Kyōto chihō rōdō undō shi hensankai, 1968.

Wetter, Gustav A. *Dialectical Materialism*. New York: Praeger, 1958.

Wilson, George M. *Radical Nationalist in Japan: Kita Ikki, 1883-1937*. Cambridge: Harvard University Press, 1969.

Wolfe, Bertram D. *Marxism, 100 Years in the Life of a Doctrine*. New York: Dial Press, 1965.

Yamaguchi-ken bunka shi 山口県文化史 (A cultural history of Yamaguchi Prefecture). Yamaguchi, 1959.

Yamanabe Kentarō 山辺健太郎. "Jūgatsu kakumei ga Nihon ni ataeta eikyō," 十月革命が日本に与えた影響 (The influence of the October Revolution on Japan), *Zen'ei* 前衛 (Vanguard), pp. 124–148 (December 1957).

Yazaki Takeo. *Social Change and the City in Japan*. Tokyo: Japan Publications, Inc., 1968.

Yoshida Kyūichi 吉田久一. "Muga ai undō to Kawakami Hajime" 無我愛運動と河上肇 (The selfless love movement and Kawakami Hajime), *Nihon rekishi* 日本歴史 (Japanese history), 165: 2–17 (March 1962).

glossary

Abe Isoo 安部磯雄
Akahata 赤旗
Akamatsu Iomaro 赤松五百麿
Amano Keitarō 天野敬太郎
Arahata Kanson 荒畑寒村
Araki Torasaburō 荒木寅三郎
Ashio 足尾
Asō Hisashi 麻生久
Bōchō Kyōiku Kai 防長教育会
Bōchō shimbun 防長新聞
bungakkai 文学会
bungakkō 文学校
bushidō 武士道
chindonya チンドン屋
Chōshū 長州
chōwa 調和
chūgakkō 中学校
dadakko 駄々子
Daitō Gitetsu 大東義徹
deshi 弟子
Doi Bansui 土井晩翠
Dōshisha 同志社
Ebina Danjō 海老名弾正
enzetsukai 演説会
Fūgetsu 楓月
fukoku kyōhei 富国強兵
Fukuda Tokuzō 福田徳三
Fukumoto Kazuo 福本和夫
Fukuzawa Yukichi 福沢諭吉
Gakuyūkai 学友会
gōdōshugi 合同主義
Gokuchū nikki 獄中日記
gōshi 郷士
Hagi 萩
haiku 俳句
hammon seinen 煩悶青年
haori 羽織
Hasegawa Nyozekan 長谷川如是閑
Heimin shimbun 平民新聞

Higuchi Ichiyō 樋口一葉
Hiramatsu 平松
hirazamurai 平侍
Hirohito 裕仁
Honda Kōzō 本田弘蔵
Hososako Kanemitsu 細迫兼光
Hozumi Yatsuka 穂積八束
Inoue Kaoru 井上馨
Inoue Kowashi 井上毅
Inoue Shie 井上シエ
Ishikawa Kōji 石川興二
Itō Hirobumi 伊藤博文
Itō Shōshin 伊藤証信
Iwakuni 岩国
Iwanami 岩波
Iwata Yoshimichi 岩田義道
Jiji shimpō 時事新報
jimbunshugi 人文主義
jindōshugi 人道主義
jingi 仁義
jimponshugi 人本主義
jinrikisha 人力車
jun-tenkōsha 準転向者
Kagekiha 過激派
Kaihō 解放
Kaihō 会報
kaisetsu 解説
kami 神
kamidana 神棚
Kanai En 金井延
Kanbe Masao 神戸正雄
Kantō 関東
Katayama Sen 片山潜
Katsura Tarō 桂太郎
Kawada Jirō 河田嗣郎
Kawakami Hajime 河上肇
Kawakami Hide 河上秀
Kawakami Iwa 河上イハ
Kawakami Kin'ichi 河上謹一

Kawakami Matasaburō 河上又三郎
Kawakami Nobusuke 河上暢輔
Kawakami Sai'ichirō 河上才一郎
Kawakami Sakyō 河上左京
Kawakami Shizuko 河上しづ子
Kawakami Sunao 河上忠
Kawakami Tazu 河上田鶴
Kawakami Yoshiko 河上芳子
Kawasaki 川崎
Kazama Jōkichi 風間丈吉
keiken-teki hōsoku 経験的法則
keikoku saimin 経国済民
keikoku saisei 経国済世
keisei 経世
keizai 経済
"Keizai-teki shikansetsu no shin kachi" 経済的史観説の新価値
Kikkawa Hiroie 吉川広家
Kinoshita Naoe 木下尚江
Kitamura Tōkoku 北村透谷
Kōbundō 弘文堂
Kōan 公案
Kokka gakkai zasshi 国家学会雑誌
kokkashugi 国家主義
Kokugo no tame ni 国語のために
kokutai 国体
Kosui Hiroto 戸水寛人
kotatsu 炬燵
kōtō-chūgakkō 高等中学校
kōtō-gakkō 高等学校
Kōtoku Shūsui 幸徳秋水
Kuroda Reiji 黒田礼二
"Kushida hōgakushi ni kotau" 櫛田法学士に答う
Kunikida Doppo 国木田独歩
Kushida Tamizō 櫛田民蔵
Kuwata Kumazō 桑田熊蔵
Kyōbundō 京文堂
kyūdōsha 求道者
Mainichi shimbun 毎日新聞
manjū 饅頭
Marukusushugi kōza マルクス主義講座
Maruyama Kanji 丸山幹治
Matsuda Masahisa 松田正久
Matsumura Noboru 松村登

Matsuzaki Kuranosuke 松崎蔵之助
Meiji 明治
Mitsui 三井
Mizutani Chōsaburō 水谷長三郎
Mori Arinori 森有礼
muga ai 無我愛
Muga En 無我苑
Namba Daisuke 難波大助
Nichū-Yūkōkai 日中友好会
Nihon gaishi 日本外史
Nihon ichi 日本一
Nihon rekishi daijiten 日本歴史大辞典
ninshikiron 認識論
Nishida Kitarō 西田幾多郎
Nishikimi 錦見
Nishiyama Yūji 西山雄次
Nō 能
nōhonshugi 農本主義
Ogawa Gōtarō 小川郷太郎
Ōhara Shakai Mondai Kenkyūsho 大原社会問題研究所
Okamura Shōichi 岡村正一
Ōkuma Nobuyuki 大熊信行
Ōmachi Keigetsu 大町桂月
Ōmori 大森
Ozaki Yukio 尾崎行雄
Rai Sanyō 頼山陽
risshin shusse 立身出世
Rōdō Nōmintō 労働農民党
Rōgakkai 労学会
Sakai Toshihiko 堺利彦
Sakisaka Itsurō 向坂逸郎
Sano Manabu 佐野学
Seijiteki Jiyū Kakutoku Rōnō Dōmei 政治的自由獲得労農同盟
"*Seisan seisaku ka bumpai seisaku ka*" 生産政策か分配政策か
Seisho no kenkyū 聖書の研究
sensei 先生
Senzan Bansuirō 千山万水楼
Shakai Kagaku Kenkyūkai 社会科学研究会
shakai seisaku 社会政策
Shakai Seisaku Gakkai 社会政策学会
Shibusawa Eiichi 渋沢栄一

Shihonron nyūmon 資本論入門
Shihonron ryakkai 資本論略解
Shimazaki Tōson 島崎藤村
Shin Rōnōtō 新労農党
Shiraishi Bon 白石凡
Shirakaba ha 白樺派
shishi 志士
shishi jinjin 志士仁人
shō-gakkō 小学校
Shōnen en 少年園
Suekawa Hiroshi 末川博
Sugamo 巣鴨
Suzuki Bunji 鈴木文治
Suzuki Yasuzō 鈴木安蔵
tachiba 立場
Taguchi Ukichi 田口卯吉
taishi ichiban 大死一番
Taishō nichinichi 大正日日
Tajima Kinji 田島錦治
Takayama Chogyū 高山樗牛
Takayama Gi'ichi 高山義一
Takekurabe たけくらべ
Tanabe Gen 田辺元
Tanaka Seigen 田中清玄
Tanaka Shōzō 田中正造
Taniguchi Zentarō 谷口善太郎
Tayama Katai 田山花袋
tenka no kōki 天下の公器
Toda Umiichi 戸田海市
Tokuda Kyūichi 徳田球一
tokushu 特殊
Tokutomi Sohō 徳富蘇峰
Tōkyō keizai zasshi 東京経済雑誌
Tōku de kasuka ni kane ga naru 遠くでかすかに鐘が鳴る
Toranomon 虎の門
Tozawa Masayasu 戸沢正保
Uchimura Kanzō 内村鑑三
waka 和歌
Wakanashū 若菜集
Yamagata Aritomo 山県有朋
Yamaguchi 山口
Yamakawa Hitoshi 山川均
Yamamoto Senji 山本宣治
Yomiuri shimbun 読売新聞
Yorozu chōhō 万朝報
Yosano Tekkan 与謝野鉄幹
Yoshida Shōin 吉田松陰
Yoshino Sakuzō 吉野作造
"Yuibutsushikan to kaikyū tōsō-setsu oyobi seitō-ha keizaigaku to no kankei" 唯物史観と階級闘争説及び正統派経済学との関係
Zenkoku Nōmin Kumiai 全国農民組合
Zenkoku Rōnō Taishūtō 全国労農大衆党

index

Abe Isoo, 21, 44
Adams, Brooks, 88
Agricultural Management in Japan (Nihon nōseigaku), 57
Akamatsu Iomaro, 133
Amano Keitarō, 197n27
Arahata Kanson, 123, 151
Araki Torasaburō, 142, 145
Asahi shimbun, 79, 87, 94, 101, 102, 147, 160, 187n41; interpretation of the Russian Revolution, 98; Osaka Asahi incident, 100, 144
Ashio Copper Mine incident, 31-32
Association for the Study of Social Policy, 53-55, 71, 96

Bentham, Jeremy, 111
Berlin, 52, 81, 124, 157
Bōchō Educational Association, 12, 14, 15
Bōchō shimbun, 31-32
Böhm-Bawerk, 66
Bolshevism, *see* Marxism
Booth, Charles, 88
Bowley, Charles, 88
Brief Explanation of Capital (Shihonron ryakkai), 154
Buddhism, 6, 10, 24-25, 26, 46-47, 78, 166, 168, 172; Zen, 28, 48, 71, 80

Capital, 101, 104, 105, 106, 145, 148, 149, 154, 156, 157-158, 166, 171, 173
Carlyle, Thomas, 91, 92, 99, 126, 155, 162
Central Forum (Chūō kōron), 143
Chamberlain, 81
Chōshū (Yamaguchi prefecture), 3, 4, 13, 18, 20, 21, 25, 29, 72, 158, 167; emphasizing education, 11-12, 16
Christian Ladies Moral Reform Society, 32
Christian socialism, *see* Socialism
Christianity, 21, 22, 24, 29-33, 36, 44, 45, 46, 47, 50, 62-63, 68, 96, 101, 110, 116, 166, 172

Comintern, 142, 151, 154, 156, 157, 159, 163
Commoner newspaper (Heimin shimbun), 45
Communism, *see* Marxism
Communist International, *see* Comintern
Communist Manifesto, 115, 119, 120, 121
Confucianism, 13, 16, 21, 22, 26, 30, 31, 47, 49, 60, 65, 69, 93, 103, 115-116, 171, 172; *Analects,* 64; *Great Learning,* 64, 89. *See also* Nationalism
Critique of Socialism (Shakaishugi hyōron), 44-45, 46, 51, 68

Daitō Gitetsu, 18
Delbert, 95
Dialectical materialism, 135-143, 155. *See also* Fukumoto Kazuo; Marxism; Economics
Doi Bansui, 15
Dōshisha University, 101

Ebina Danjō, 45
Economic determinism, *see* Historical materialism
Economics and Human Existence (Keizai to jinsei), 67, 71
Economics: as a moral vocation, 36-42, 165; as social policy, 52-55, 165; conflict between ethics and economics, 55-64, 165; non-Marxist economic theory, 87-89, 91-92, 111-112, 126; Marxist economic theory, 106-113, 115-116, 117-128, 135-139, 143, 155-156, 162, 165, 168
Engels, 119, 120-121, 124, 137
England, 43, 78, 79-80, 81, 82, 83, 87, 88, 89, 91, 103; English language, 12, 80, 88, 91; English economists, 52-53, 76, 91, 105, 106

Fisher, Irving, 66

217

Friendly Society: Kyoto, 100-101; Kansai Federation, 132
Fukoku kyōhei, 4, 14, 59
Fukuda Tokuzō, 96
Fukui Takahara (Kōji), 132
Fukumoto Kazuo, 123, 142, 143, 173; criticism of Kawakami, 135-137, 139-141, 148
Fukuzawa Yukichi, 14, 25; *Current Affairs (Jiji shimpō)*, 13
Fundamental Principles of Political Economics (Keizaigaku jō no kompon kannen), 42

Gakuyūkai, 141
Garden of Selflessness (Muga En), 46-51, 57, 95, 97, 138, 150, 158, 169
German historical school, *see* Germany
Germany, 52, 53, 76, 79, 83, 87, 88, 91, 103, 116, 124, 135; German language, 13, 18, 77, 94, 133, 157; German texts, 27, 103, 109, 132; German economists, 37, 40, 41, 42, 52-53, 72, 76, 105, 118, 119, 124, 138, 165; German culture/philosophy, 81, 106, 116, 118, 127, 139
Gershuni, Gregory, 122
Godwin, Thomas, 95

Hasegawa Nyozekan, 102. *See also We*
Hegel, 76, 106, 108, 116, 127
Higuchi Ichiyō, 15
Hiramatsu, *see* Kazama Jōkichi
Historical Development of Capitalist Economics (Shihonshugi keizaigaku no shiteki hatten), 126, 155
Historical materialism, 39-42, 69, 102, 106-111, 113, 115, 116, 117, 118-128, 133, 137, 139, 142-143, 155, 162
History of Modern Economic Thought (Kinsei keizai shisō shi ron), 115
Honda Kōzō, *see* Kawakami Hajime
Horie Muraichi, 100, 132, 197n27
Hososako Kanemitsu, 151-154
Hozumi Yatsuka, 21
Humanism, 37, 45, 46, 58, 59, 63, 67-69, 88, 91, 92, 93, 104, 112-116, 128, 131, 189n32

Inoue Kowashi, 16
Ishikawa Kōji, 133

Itō Hirobumi, 4, 8, 17. *See also* Meiji oligarchs
Itō Shōshin, 46, 47, 48, 50. *See also* Garden of Selflessness
Iwata Yoshimichi, 132, 141, 159, 160, 195n56

Japan Socialist Party, 103
Japanese Communist Party, 135, 141, 142, 144, 151, 154-162, 164, 173
Japanese Marxism, *see* Marxism

Kanai En, 27, 37, 52, 53, 66. *See also* Association for the Study of Social Policy
Kanbe Masao, 66
Kant, 76, 116
Katayama Sen, 21, 35-36, 55, 68, 89, 198n27
Kautsky, 119, 124
Kawada Jirō, 66
Kawakami Hajime: early childhood, 3, 4-11; religious influences, 6, 10, 16, 17, 24-25, 26, 29-31, 32, 33, 45, 47, 49, 61-64, 69, 90, 92, 110-111, 113, 115, 116, 118-119, 122-123, 127, 131, 139-140, 149, 165, 172-173; health, 8-9, 48, 116, 117, 136, 140, 160, 168; character, 9-11, 13-14, 15, 16, 18, 24, 25, 30, 34, 37, 38, 42, 43, 48-49, 50, 59, 63-64, 67-69, 76, 88-90, 99, 109, 113, 119, 139, 148, 150, 155, 162, 165, 167, 169-173; education, 12-16, 18, 23, 24, 25-28, 34, 36, 37, 52, 68, 71, 138, 169; nationalistic thought, 13-14, 16-17, 59-64, 65, 67, 69, 72-75, 77, 91, 93, 169; academic interests, 12, 15-17, 24, 26, 27, 36-40, 41, 42-43, 46, 50, 51, 55-60, 62-64, 66, 67, 76, 83, 88-93, 104, 109, 111, 113, 117-127, 131-134, 136-143, 147, 148, 154-155, 165, 172-173; marriage, 34; political interests, 36, 39, 44-47, 51, 55, 66, 68-70, 72, 82, 83, 87, 91, 92, 93, 94, 96, 97-105, 109-110, 111, 113, 114, 118-128, 131-145, 147-173; concern for agriculture, 56-60; espousing social harmony, 60-64, 67, 69, 72-73, 93, 98; studies on Japanese culture, 72-76, 79-81; abroad, 76-83, 87; rejection of Malthus and Smith, 89-90, 91, 112; later career, 94, 104, 117, 131-133, 136, 139-141, 145;

involvement in the proletarian movement, 100-101, 132-134, 143-144, 147, 154, 156, 169, 171
Kawakami Hajime Commemorative Society, 170
Kawakami Hide (wife): family background, 34; marriage to Hajime, 34; running the household, 41, 50, 62, 77, 146, 153, 196n4; caring for sick son, 71, 134; loyalty to Hajime, 149, 157-158, 167
Kawakami Iwa (grandmother), 5-7, 14, 20
Kawakami Kin'ichi (uncle), 22-23, 32, 167
Kawakami Masao (son), 71, 134
Kawakami Nobusuke (stepbrother), 5, 32
Kawakami Sakyō (brother), 5, 10, 14, 167
Kawakami Shizuko (daughter), 11, 196n4
Kawakami Sunao (father), 3-5, 7, 9-10, 11, 14, 18
Kawakami Tazu (mother), 5-6, 32, 167
Kawakami Yoshiko (daughter), 156, 159, 195n56, 196n4
Kazama Jōkichi, 159-160
Kikkawa family, lords of Iwakuni, 3, 10, 12, 34
Kunikida Doppo, 15
King, 88
Kinoshita Naoe, 21, 24, 32, 36, 68, 92. *See also* Christianity
Kitamura Tōkoku, 14
Kobayashi Terutsugi, 100, 132
Kojima Yūma, 102
kokutai (national polity), 68, 73
Kosui Hirōto, 68
Kōtoku Shūsui, 44, 96
Kumazawa Banzan, 63
Kushida Tamizō, 102, 115, 119, 147-148, 149, 152; student of Kawakami, 87, 94, 133; study abroad, 124; criticism of Kawakami, 94-97, 99, 105, 115-116, 126-128, 131, 141, 143, 155; correspondence with Kawakami, 100, 117, 136, 140
Kuwata Kumazō, 53
Kyoto Imperial University, 64-65, 66, 83, 102, 144; Kawakami at, 65, 66, 80, 97, 104, 111, 117, 131-133, 145, 171; students of, 94, 100, 141-142, 145, 150

Labor-Farmer faction (Rōnōha), 151

Labor-Farmer Party, 143-145, 149, 150-151. *See also* New Labor-Farmer Party
Labor movement, 21, 54-55, 61-62, 99-101, 132-133; proletarian party movement, 143-145, 147-154, 156, 169. *See also* Friendly Society; Labor-Farmer Party
Lectures on Marxism (Marukusushugi kōza), 144
Leisurely Meditation in Prison (Gokuchū zeigo), 168
Lenin, 118, 123, 124, 125, 126, 136, 137-138, 142, 148, 158, 162, 191n21, 191n24, 191n29
Leninism, *see* Marxism
Li Ta-chao, 190n5
Liang Ch'i-ch'ao, 82
Liberation (Kaihō), 96, 114
Literary World (Bungakkai), 15
London, 79-80, 82, 83

Malthus, 35, 61, 89, 105, 112
Marx, Karl, *see* Marxism
Marxism, 38, 39, 41, 66, 68, 69, 76, 94-95, 96, 97, 101, 102, 103-110, 111, 112, 113-116, 117, 118-125, 126, 127, 128, 131-133, 135-143, 144, 145, 147-173
Matsuda Masahisa, 18
Matsukata Saburō, 132
Matsumura Noboru, 159-160
Matsuzaki Kuranosuke, 27, 36, 41, 42, 51, 66
Maruyama Kanji, *Taisho Daily (Taishō nichinichi)*, 102
Meiji Japan: abolition of feudalism, 11, 56; intellectual ferment, 11-13, 15, 17, 21, 23, 25, 40; agonized youth, 28-29, 45. *See also* Socialism
Meiji oligarchs, 4, 20, 22, 25, 72
Meiji Restoration, 3, 17, 34
Mill, John Stuart, 111, 112, 126, 128
Miyakawa Minoru, 132, 154, 197n27
Mizuta Mikio, 194n20
Mizutani Chōsaburō, 132, 144, 151, 153
Mōri (Chōshū), 3
Mori Arinori, 7

Nabeyama Sadachika, 163
Namba Daisuke, 134-135
National Academic Society magazine (Kokka gakkai zasshi), 42, 52
Nationalism, 3, 4, 13-14, 16-17, 20, 40, 53, 59-60, 63-64, 65, 67, 68,

69, 72-73, 75, 77, 79, 82, 91, 93, 169
Natsume Sōseki, 8, 25, 78-79
Neo-Confucianism, see Confucianism; Tokugawa Neo-Confucianism
New Labor-Farmer Party, 149, 153-154, 156, 171
New Magazine of Japanese Economics (Nihon keizai shinshi), 51, 55, 57
Nishida Kitarō, 139
Nishikimi, 3, 5
Nishiyama Yūji, 197n26
Nōhonshugi, 57-59

Ogawa Gōtarō, 66
Ohara Research Institute, 115, 131, 133-134
Okamura Shōichi, 29
Ōkōchi Kazuo, 170
Ōkuma Nobuyuki, 197n27
Ōmachi Keigetsu, 16
Ōmori Gang affair, see Ōtsuka Yūshō
On the Reverence for Agriculture in Japan (Nihon sonnō ron), 42, 57
Ōtsuka Yūshō (Kawakami's brother-in-law), 10, 151-153, 156-160, 164, 196n8
Ōuchi Hyōe, 27, 92, 197n26
Oyama Ikuo: receiving Kawakami's aid, 144, 145, 147, 151, 152, split with Kawakami, 153-154, 156, 170
Ozaki Yukio, 18

Paris, 77-78
Peace Preservation Act, 141, 142, 164
Pierson, N. G., 66
Plekhanov, 118

Rai San'yō, 13
Reconstruction (Kaizō), 103, 133
Red Flag (Akahata), 157-158, 173
Reflections on Our Homeland (Sokoku o kaerimite), 79, 82
Report (Kaihō), 13
Research in Social Problems (Shakai mondai kenkyū), 102, 103, 109, 114, 117-118, 125, 147, 149
Research on Historical Materialism (Yuibutsu shikan kenkyū), 118
Revolution, 17, 59, 62, 95, 101, 108, 115, 134, 135, 137, 139, 142, 148, 156-157, 162, 163, 168, 171, 173; Russian Revolution, 98-99, 122-123, 124-126, 127, 131, 133, 134; French Revolution, 95; Kawakami's discussion of, 118-122, 124-126

Ricardo, 105, 112
Rice Riots, 98, 99-100
Risshin shusse, 13, 47
Ruskin, James, 91, 92, 99, 111, 112, 118, 126, 155
Russia (Soviet Union), 68, 71, 79, 95, 98, 99, 124, 125, 126, 134, 137, 139, 149
Russian Revolution, see Revolution
Russo-Japanese War, 51, 54, 56, 71
Ryazanov, 149

Sakai Toshihiko, 103, 104, 113-115, 123, 151
Sakisaka Itsurō, 93, 94
Samurai spirit, 5, 6, 7, 10, 30, 94, 167, 173
Sano Manabu, 156, 163
Satō Nobuhiro, 68
Satsuma, 4, 20, 72
Selfless Love Movement, see Garden of Selflessness
Seligman, Edwin R. A., 38-39, 41, 51
Senzen Bansuirō, see Kawakami Hajime
Shaw, Bernard, 80
Shibusawa Eiichi, 61, 64
Shiga Yoshio, 173
Shimazaki Tōson, 15, 29, 77-78, 198n32; *Young Greens (Wakanashū)*, 29, 50
Shinto, 6, 17, 26, 136
Shiraishi Bon, 100, 132, 196n4, 197n26, 197n27
Shishi (samurai patriots), 3, 167-168; shishi spirit, 4, 13, 17, 25, 28, 30, 31-32, 98, 168, 172
Smart, Thomas, 91, 118, 155
Smith, Adam, 60, 89, 91, 97, 105, 111-112
Social Darwinism, 41, 81
Social Organization and Social Revolution (Shakai soshiki to shakai kakumei), 118
Social policy, 52-55, 64, 66, 67, 69-70, 96, 99, 126, 165. See also Germany; Association for the Study of Social Policy
Social Science Study Club (Shakai kagaku kenkyūkai), 141, 142, 145
Socialism, 31, 36, 38, 44, 51, 52, 54, 62, 68, 69, 70, 91, 93, 96, 100, 101, 112, 120, 122, 135, 151; European, 20, 53; Christian socialism, 21-22, 36, 44-45, 96, 110; Utopian socialism, 109-110; Marxian socialism, 94, 103, 107, 109-110, 113, 114, 118, 123-126

Soviet Union, *see* Russia
Statecraft, 4, 12, 13, 17, 18, 37, 53, 165
Stevens, Herbert, 12
Studies in Economics (*Keizaigaku kenkyū*), 71
Suzuki Bunji, 99, 100-101
Suzuki Shigetoshi, 196n4, 196n8
Suzuki Yasuzō, 150, 167

Taguchi Ukichi, 52, 65
Tajima Kinji, 66
Takayama Chogyū, 29
Takayama Gi'ichi, 100-101, 132
Takeda Akira (Sei), 77, 82
Tale of Genji, 15
Tale of Poverty (*Bimbō monogatari*), 87-94, 95, 96, 98, 99, 101, 118, 131, 155
Tanabe Gen, 139
Tanaka Seigen, 131
Tanaka Shōzō, 21, 26, 31, 32, 92, 171
Taniguchi Zentarō, 197n27
Tayama Katai, 15
Tenkō, 163
Theories of Political Economics (*Keizaigaku genron*), 42
Toda Umiichi, 66
Tokuda Kyūichi, 173
Tokugawa, 3, 4, 19, 56, 62, 64, 90, 165
Tokugawa neo-Confucianism, 62-64, 165
Tokutomi Sohō, 17
Tokyo, 13, 22, 34, 46, 50, 65, 71, 143, 147, 149, 151-153, 158, 160; turn-of-the-century, 19-22
Tokyo Imperial University, 14, 27-28, 65, 152; Kawakami at, 18, 20, 23, 36; faculty of, 21, 26-27, 37, 44, 52, 53, 68, 115; students of, 22, 23, 31, 131, 132, 140
Tolstoy, Leo, 44-45, 46, 48, 50, 198n32
Toranomon incident, *see* Namba Daisuke
Torii Sosen, 100
Toynbee, Arnold, 43, 50
Tozawa Shōzō, 82
Trend of the Times (*Jisei no hen*), 71
Trotsky, 118, 124, 125, 126

Uchimura Kanzō, 21, 24, 25, 26, 29, 44
United States, 58, 87, 88, 103, 122
Universal Manhood Suffrage Act, 144

Vanguard (*Zen'ei*), 123
Views on Social Problems (*Shakai mondai kanken*), 99, 118, 143

Wagner, Roger, 52
Watsuji Tetsurō, 72, 142
We (*Warera*), 102, 103, 144, 147
Worker-Student Society (Rōgakkai), 100, 101, 132

Yamagata Aritomo, 4, 7, 17
Yamaguchi City, 14, 17, 26
Yamakawa Hitoshi, 103, 104, 123, 151
Yamamoto Senji, 144, 151-153
Yomiuri newspaper, 15, 44, 46, 50, 51
Yosano Tekkan, 15
Yoshida Shōin: as a shishi, 13, 17, 172; inspiring Kawakami, 18, 31-32, 45, 167, 171
Yoshino Sakuzō, 42, 51, 187n41
Yūaikai, *see* Friendly Society

harvard east asian series

1. *China's Early Industrialization: Sheng Hsuan-huai (1884–1916) and Mandarin Enterprise.* By Albert Feuerwerker.
2. *Intellectual Trends in the Ch'ing Period.* By Liang Ch'i-ch'ao. Translated by Immanuel C. Y. Hsü.
3. *Reform in Sung China: Wang An-shih (1021–1086) and His New Policies.* By James T. C. Liu.
4. *Studies on the Population of China, 1368–1953.* By Ping-ti Ho.
5. *China's Entrance into the Family of Nations: The Diplomatic Phase, 1858–1880.* By Immanuel C. Y. Hsü.
6. *The May Fourth Movement: Intellectual Revolution in Modern China.* By Chow Tse-tsung.
7. *Ch'ing Administrative Terms: A Translation of the Terminology of the Six Boards with Explanatory Notes.* Translated and edited by E-tu Zen Sun.
8. *Anglo-American Steamship Rivalry in China, 1862–1874.* By Kwang-Ching Liu.
9. *Local Government in China under the Ch'ing.* By T'ung-tsu Ch'ü.
10. *Communist China, 1955–1959: Policy Documents with Analysis.* With a foreword by Robert R. Bowie and John K. Fairbank. (Prepared at Harvard University under the joint auspices of the Center for International Affairs and the East Asian Research Center.)
11. *China and Christianity: The Missionary Movement and the Growth of Chinese Antiforeignism, 1860–1870.* By Paul A. Cohen.
12. *China and the Helping Hand, 1937–1945.* By Arthur N. Young.
13. *Research Guide to the May Fourth Movement: Intellectual Revolution in Modern China, 1915–1924.* By Chow Tse-tsung.
14. *The United States and the Far Eastern Crisis of 1933–1938: From the Manchurian Incident through the Initial Stage of the Undeclared Sino-Japanese War.* By Dorothy Borg.
15. *China and the West, 1858–1861: The Origins of the Tsungli Yamen.* By Masataka Banno.
16. *In Search of Wealth and Power: Yen Fu and the West.* By Benjamin Schwartz.
17. *The Origins of Entrepreneurship in Meiji Japan.* By Johannes Hirschmeier, S.V.D.
18. *Commissioner Lin and the Opium War.* By Hsin-pao Chang.
19. *Money and Monetary Policy in China, 1845–1895.* By Frank H. H. King.
20. *China's Wartime Finance and Inflation, 1937–1945.* By Arthur N. Young.
21. *Foreign Investment and Economic Development in China, 1840–1937.* By Chiming Hou.

22. *After Imperialism: The Search for a New Order in the Far East, 1921-1931.* By Akira Iriye.
23. *Foundations of Constitutional Government in Modern Japan, 1868-1900.* By George Akita.
24. *Political Thought in Early Meiji Japan, 1868-1889.* By Joseph Pittau, S.J.
25. *China's Struggle for Naval Development, 1839-1895.* By John L. Rawlinson.
26. *The Practice of Buddhism in China, 1900-1950.* By Holmes Welch.
27. *Li Ta-chao and the Origins of Chinese Marxism.* By Maurice Meisner.
28. *Pa Chin and His Writings: Chinese Youth Between the Two Revolutions.* By Olga Lang.
29. *Literary Dissent in Communist China.* By Merle Goldman.
30. *Politics in the Tokugawa Bakufu, 1600-1843.* By Conrad Totman.
31. *Hara Kei in the Politics of Compromise, 1905-1915.* By Tetsuo Najita.
32. *The Chinese World Order: Traditional China's Foreign Relations.* Edited by John K. Fairbank.
33. *The Buddhist Revival in China.* By Holmes Welch.
34. *Traditional Medicine in Modern China: Science, Nationalism, and the Tensions of Cultural Change.* By Ralph C. Croizier.
35. *Party Rivalry and Political Change in Taishō Japan.* By Peter Duus.
36. *The Rhetoric of Empire: American China Policy, 1895-1901.* By Marilyn B. Young.
37. *Radical Nationalist in Japan: Kita Ikki, 1883-1937.* By George M. Wilson.
38. *While China Faced West: American Reformers in Nationalist China, 1928-1937.* By James C. Thomson, Jr.
39. *The Failure of Freedom: A Portrait of Modern Japanese Intellectuals.* By Tatsuo Arima.
40. *Asian Ideas of East and West: Tagore and His Critics in Japan, China, and India.* By Stephen N. Hay.
41. *Canton under Communism: Programs and Politics in a Provincial Capital, 1949-1968.* By Ezra F. Vogel.
42. *Ting Wen-chiang: Science and China's New Culture.* By Charlotte Furth.
43. *The Manchurian Frontier in Ch'ing History.* By Robert H. G. Lee.
44. *Motoori Norinaga, 1730-1801.* By Shigeru Matsumoto.
45. *The Comprador in Nineteenth Century China: Bridge between East and West.* By Yen-p'ing Hao.

46. *Hu Shih and the Chinese Renaissance: Liberalism in the Chinese Revolution, 1917-1937.* By Jerome B. Grieder.
47. *The Chinese Peasant Economy: Agricultural Development in Hopei and Shantung, 1890-1949.* By Ramon H. Myers.
48. *Japanese Tradition and Western Law: Emperor, State, and Law in the Thought of Hozumi Yatsuka.* By Richard H. Minear.
49. *Rebellion and Its Enemies in Late Imperial China: Militarization and Social Structure, 1796-1864.* By Philip A. Kuhn.
50. *Early Chinese Revolutionaries: Radical Intellectuals in Shanghai and Chekiang, 1902-1911.* By Mary Backus Rankin.
51. *Communications and Imperial Control in China: Evolution of the Palace Memorial System, 1693-1735.* By Silas H. L. Wu.
52. *Vietnam and the Chinese Model: A Comparative Study of Nguyen and Ch'ing Civil Government in the First Half of the Nineteenth Century.* By Alexander Barton Woodside.
53. *The Modernization of the Chinese Salt Administration, 1900-1920.* By S. A. M. Adshead.
54. *Chang Chih-tung and Educational Reform in China.* By William Ayers.
55. *Kuo Mo-jo: The Early Years.* By David Tod Roy.
56. *Social Reformers in Urban China: The Chinese Y.M.C.A., 1895-1926.* By Shirley S. Garrett.
57. *Biographic Dictionary of Chinese Communism, 1921-1965.* By Donald W. Klein and Anne B. Clark.
58. *Imperialism and Chinese Nationalism: Germany in Shantung.* By John E. Shrecker.
59. *Monarchy in the Emperor's Eyes: Image and Reality in the Ch'ien-lung Reign.* By Harold L. Kahn.
60. *Yamagata Aritomo in the Rise of Modern Japan, 1838-1922.* By Roger F. Hackett.
61. *Russia and China: Their Diplomatic Relations to 1728.* By Mark Mancall.
62. *The Yenan Way in Revolutionary China.* By Mark Selden.
63. *The Mississippi Chinese: Between Black and White.* By James W. Loewen.
64. *Lang Ch'i-ch'ao and Intellectual Transition in China, 1890-1907.* By Hao Chang.
65. *A Korean Village: Between Farm and Sea.* By Vincent S. R. Brandt.
66. *Agricultural Change and the Peasant Economy of South China.* By Evelyn S. Rawski.
67. *The Peace Conspiracy: Wang Ching-wei and the China War, 1937-1941.* By Gerald Bunker.

68. *Mori Arinori.* By Ivan Hall.
69. *Buddhism under Mao.* By Holmes Welch.
70. *Student Radicals in Prewar Japan.* By Henry Smith.
71. *The Romantic Generation of Modern Chinese Writers.* By Leo Ou-fan Lee.
72. *Deus Destroyed: The Image of Christianity in Early Modern Japan.* By George Elison.
73. *Land Taxation in Imperial China, 1750-1911.* By Yeh-chien Wang.
74. *Chinese Ways in Warfare.* Edited by Frank A. Kierman Jr. and John K. Fairbank.
75. *Pepper, Guns, and Parleys: The Dutch East India Company and China, 1662-1681.* By John E. Wills Jr.
76. *A Study of Samurai Income and Entrepreneurship: Quantitative Analyses of Economic and Social Aspects of the Samurai in Tokugawa and Meiji Japan.* By Kozo Yamamura.
77. *Between Tradition and Modernity: Wang T'ao and Reform in Late Ch'ing China.* By Paul A. Cohen.
78. *The Abortive Revolution: China under Nationalist Rule, 1927-1937.* By Lloyd E. Eastman.
79. *Russia and the Roots of the Chinese Revolution, 1896-1911.* By Don C. Price.
80. *Toward Industrial Democracy: Management and Workers in Modern Japan.* By Kunio Odaka.
81. *China's Republican Revolution: The Case of Kwangtung, 1895-1913.* By Edward J. M. Rhoads.
82. *Politics and Policy in Traditional Korea.* By James B. Palais.
83. *Folk Buddhist Religion: Dissenting Sects in Late Traditional China.* By Daniel L. Overmyer.
84. *The Limits of Change: Essays on Conservative Alternatives in Republican China.* Edited by Charlotte Furth.
85. *Yenching University and Sino-Western Relations, 1916-1952.* By Philip West.
86. *Japanese Marxist: A Portrait of Kawakami Hajime, 1879-1946.* By Gail Lee Bernstein.